Fundamentals of

Geographic Information System

Fundamentals of Geographic Information System

Debashis Chakraborty
Rabi N. Sahoo
Scientists
Division of Agricultural Physics
Indian Agricultural Research Institute (IARI), New Delhi

Viva Books
New Delhi | Mumbai | Chennai | Kolkata | Bengaluru | Hyderabad | Kochi | Guwahati

Publisher's note

First Published 2007
Reprinted 2010, 2015, 2016, 2017, 2019, 2020, 2023, 2024, 2025, 2026

Viva Books Private Limited

- 4737/23, Ansari Road, Daryaganj, New Delhi 110 002
  Tel. 42242200, 23258325, 23283121, Email: vivadelhi@vivagroupindia.net
- 76, Service Industries, Shirvane, Sector 1, Nerul, Navi Mumbai 400 706
  Tel. 27721273, 27721274, Email: vivamumbai@vivagroupindia.net
- Megh Towers, Old No. 307, New No. 165, Poonamallee High Road, Maduravoyal
  Chennai 600 095 Tel. 23780991, 23780992, Email: vivachennai@vivagroupindia.net
- B-103, Jindal Towers, 21/1A/3 Darga Road, Kolkata 700 017
  Tel. 22816713, Email: vivakolkata@vivagroupindia.net
- 7, GF, Sovereign Park Aptts., 56-58, K. R. Road, Basavanagudi, Bengaluru 560 004
  Tel. 26607409, Email: vivabangalore@vivagroupindia.net
- 101-102, Moghal Marc Apartments, 3-4-637 to 641, Narayanguda, Hyderabad 500 029
  Tel. 27564481, Email: vivahyderabad@vivagroupindia.net
- First Floor, Beevi Towers, SRM Road, Kaloor, Kochi 682 018
  Tel. 0484-2403055, 2403056, Email: vivakochi@vivagroupindia.net
- 232, GNB Road, Beside UCO Bank, Silpukhuri, Guwahati 781 003
  Tel. 0361-2666386, Email: vivaguwahati@vivagroupindia.net

www.vivagroupindia.com

ISBN: 978-81-309-0041-4

Published by Vinod Vasishtha for Viva Books Private Limited, 4737/23 Ansari Road, Daryaganj, New Delhi 110 002.

Printed and bound by Replika Press Pvt. Ltd.

CONTENTS

There's a silent revolution going on. It's a revolution that impacts each of our lives, although few of us have heard anything about it. ... The basis for this silent revolution is a technology called GIS—an acronym for Geographic Information System."—Business Week, July, 1991

Preface

With a successfully running space programme for national developments under the flagship of Department of Space (DOS) and a fast-expanding growth of the Information and Communication Technology sector, India certainly demands an edge in the global competitive market. These are two of the major pillars of economic development in the country. Unless Geoinformation Science and Technologies (encompassing Geographic Information System, Remote Sensing, Global Positioning System) are developed and applied in India with its fullest potential, the maximum benefits accrued from these core sectors will remain largely untapped. Geographic Information Science and Technology is relevant to any research and developmental organization dealing with spatial information for monitoring, policy, planning or management purposes. Introduced and known as a package or system, Geographic Information System (GIS) is undoubtedly the most sought after tool from the national to the local (municipality) level, for Schools, Universities and Research Institutions, for the private sector and also for non-governmental organizations (NGOs).

GIS needs multidisciplinary approach and its applications cut across the boundaries between various disciplines. Of late there has been inclusion of a number of courses in GIS at various undergraduate and postgraduate levels of teaching and training programmes. Keeping this in view the book is intended to be a basic level textbook, on the one hand, for the beginners and, on the other hand, as a reference book for GIS practitioners who use the system for their professional work. A well-organized textbook is very important for teaching, learning and training of successful GIS, this book tries to integrate the basic idea that GIS is for everyone.

The book is organized into 10 Chapters. An introduction to GIS and its various components and functionalities are covered in Chapter 1. This Chapter provides a conceptual and historical framework of GIS with a comprehensive review of its development as a science. Some major areas of applications are given with their possible limitations. Chapter 2 starts with the understanding of cartography as the root of GIS and continues the journey through overview of geometric models of the earth, coordinate systems and principles of map projection systems.

How maps can be represented as numbers (digits) in a GIS is explained in Chapter 3 with basic understanding of spatial data models namely, raster and vector. Chapter 4 provides an overview of the data input processes with a detailed account of Global Positioning System (GPS) as a new input data source. Chapter 5 deals with the spatial and attribute data management and data modelling aspects. Spatial analysis, the most important function that makes GIS distinct from any other mapping and data analysis system is discussed in Chapter 6, whereas Chapter 7 specifically deals with spatial interpolation and surface modeling processes with a brief introduction of the concept of artificial neural network systems. Various data output mode and devices are summarized in Chapter 8. GIS is often seen as add-on to the remote sensing systems, another modern tool of information capture about the objects and a whole chapter (Chapter 9) is devoted to their potential integration. The book ends with Chapter 10 with critical discussion of some present day key socio-technical issues involving GIS. Some relevant details as supplementary materials are provided in Boxes for interested readers and also to allow first reading without the loss of continuity. The Study Questions are given at the end for some practice. This book also contains a complete Glossary of terms, the Abbreviations and the Acronyms frequently used in GIS.

We are sure that this book will provide a step-by-step appreciation of Geographic Information Science as a discipline and the Geographic Information System as a tool to understand the real world to everyone who strives to change the world in a better and more meaningful way. Valuable comments and suggestions are solicited from the readers (debashis@iari.res.in or rnsahoo@iari.res.in).

Deabshis Chakraborty
Rabi N. Sahoo

Acknowledgements

We express our sincere thanks to the teachers who taught us the subject during our academic programs and also encouraged us to take up research activities simultaneously. We are extremely grateful to the authors (cited in Bibliography) of various publications, for the information which has enriched the quality of the book. We are grateful to Ms Bhavna of Viva Books Private Limited for their faith and perseverance in encouraging us to write the book. We are also grateful to our colleagues in the Division of Agricultural Physics, Indian Agricultural Research Institute for their constant encouragement and help. Finally, we appreciate the love and support of our dear family members during the chore of the project.

Acronyms

ACSM	American Congress on Surveying and Mapping
AGI	Association for Geographic Information
AGIS	Automated Geographic Information System
AI	Artificial/Amplified Intelligence
AM/FM	Automatic Mapping / Facilities Management
AML	Arc Macro Language
ANN	Artificial Neural Network
ANSI	American National Standards Institute
ASCII	American Standard Code for Information Interchange
ASLV	Augmented Satellite Launch Vehicle
AVL	Automatic Vehicle Location
BIOS	Basic Input Output System
BIT	Binary Digit
BLOB	Binary Large Object
BLPU	Basic Land and Property Unit
BM	Bench Mark
BMP	BitMaP
BSI	British Standards Institute
CAC	Computer Assisted Cartography
CAD	Computer Assisted (or Aided) Drafting, Design or Delivery
CAL	Computer Assisted (or Aided) Learning
CAM	Computer Assisted (or Aided) Mapping/Manufacturing
CASE	Computer Assisted (or Aided) Software Engineering

CCT	Computer Compatible Tape
CD	Compact Disk
CD-ROM	Compact Disk-Read Only Memory
CGI	Common Gateway Interface
CLIENT	A software application that interacts with a server computer
COBOL	Common Business Oriented Language
COGO	Coordinate Geometry
COORDS	x,y Coordinates (or x,y,z for three-dimensional models)
CPU	Central Processing Unit
CRT	Cathode Ray Tube
DBMS	Database Management System
DCM	Digital Cartographic Model
DDE	Dynamic Data Exchange
DEI	Data Extraction Interface
DEM	Digital Elevation Model
DIR	DIRectory
DLG	Digital Line Graph
DLL	Dynamically Linked Library
DLM	Digital Landscape Model
DN	Digital Number
DOS	Disk Operating System; Department of Space (Government of India)
DPI	Dot Per Inch
DTM	Digital Terrain Model
EDC	EROS Data Centre
EIA	Environmental Impact Assessment
EIS	Environmental Impact Statement
EOS	Earth Observation Satellite
EPA	Environmental Protection Agency
EPS	Encapsulated PostScript
ES	Expert System
ESRI	Environmental Systems Research Institute (developers of Arc/Info)
EXE	EXEcutable file
FCC	False Colour Composite
FORTRAN	FORmula TRANslation (programming language)
FTP	File Transfer Protocol
GB	GigaByte

GCP	Ground Control Point
GIF	Graphics Interchange Format
GIS	Geographic Information System
GOES	Geostationary Operational Environmental Satellite
GPS	Global Positioning System
GRAM	Geo-Referenced Area Management
GSLV	Geostationary Satellite Launch Vehicle
GUI	Graphical User Interface
HP	Hewlett-Packard
HTML	Hyper Text Markup Language
HTTP	Hyper Text Transport Protocol
HW	HardWare
I/O	Input/Output
IARI	Indian Agricultural Research Institute
IE	Internet Explorer (Microsoft browser)
IEEE	Institute of Electrical and Electronics Engineers
IIT	Indian Institute of Technology
IMD	India Meteorological Department
IMS	Information Management System
IMSD	Integrated Mission for Sustainable Development
INSAT	Indian National Satellite
IP	Internet Protocol
IRS	Indian Remote Sensing Satellite
ISDN	Integrated Services Digital Network
ISO	International Standards Institute
ISPRS	International Society of Photogrammetry and Remote Sensing
ISRO	Indian Space Research Organization
JERS	Japan Earth Resources Satellite
KB	Kilo Byte
LAN	Local Area Network
LIS	Land Information System
LUT	Look Up Table
MB	Mega Byte
MCF	Master Control Facility
MHz	Mega Hertz
MIMO	Map-In-Map-Out

MIS	Management Information System
MO	Map Object
MODEM	Modulator-DEModulator
MOIMS	Map Objects Internet Map Server
NADAMS	National Agricultural Drought Assessment and Monitoring System
NASA	National Aeronautics and Space Administration
NBSS&LUP	National Bureau of Soil Survey and Land Use Planning
NC	Network Computer
NDC	National Data Centre
NDVI	Normalized Difference Vegetation Index
NET	NETwork
NFS	Network File System
NG	National Grid
NIC	Network Interface Card
NICNET	National Informatics Centre Network
NNRMS	National Natural Resource Management System
NOAA	National Oceanographic and Atmospheric Administration
NRIS	National Resource Information System
NRSA	National Remote Sensing Agency
NSF	(US) National Science Foundation
NT	New Technology (in Windows NT)
NTF	(UK) National Transfer Format
OCR	Optical Character Recognition
ODBC	Open DataBase Connectivity
OGIS	Open Geodata Interoperability Specification
OLE	Object Linking and Embedding
OLTP	OnLine Transaction Processing
OO	Object Oriented
OODBMS	Object Oriented Data Base Management System
OS	Operating System, Ordnance Survey
PC	Personal Computer
PERL	Practical Report and Extraction Language (in UNIX)
PIXEL	Picture Element
PS	PostScript
PSLV	Polar Satellite Launch Vehicle
RAM	Random Access Memory

RDBMS	Relational DataBase Management System
RF	Representative Fraction
RGB	Red, Green, Blue
RGS	Royal Geographical Society
RIMS	Road Information Management System
RRSSC	Regional Remote Sensing Service Centre (ISRO/DOS)
SAC	Space Application Centre, DOS (Government of India)
SAS	Statistical Analysis System
SCSI	Small Computer System Interface
SDE	Spatial Database Engine
SIMS	Survey Information Management System
SIS	Spatial Information System
SPOT	System Probatoire de Observation de la Terre
SPSS	Statistical Package for the Social Sciences
SQL	Structured Query Language
SVGA	Super Video Graphics Adapter
SW	SoftWare
TC	True Colour
TCP/IP	Transport Control Protocol / Internet Protocol
TELNET	TELetype NETwork
TIFF	Tagged Image File Format
TIGER	Topologically Integrated Geographic Encoding and Referencing
TM	Thematic Map(ping)
UI	User Interface
URL	Uniform Resource Locator
UTM	Universal Transverse Mercator
VB	Visual Basic
VGA	Video Graphics Adapter
VR	Virtual Reality
VRML	Virtual Reality Modelling Language
WAN	Wide Area Network
WMO	World Meteorological Organization
WGS	World Geodetic System
WRS	World Reference System
WWW	World Wide Web

CHAPTER 1

Fundamentals of GIS

"If you will have a young man to put his travel into a little room, and in short time to gather much, this you must do; ... let him carry with him also some map or book describing the country where he travelleth, which will be a good key to his inquiry..."

Sir Francis Bacon, Essay #18, Of Travel, 1625

1.1 THE NATURE OF GEOGRAPHICAL PHENOMENA

Duane F. Marble, the then Professor of Geography at Ohio State University, USA, put a quote, made sometime by Albert Einstein, on his office door in response to the cool welcome received by the Geography Department which had taken over part of the physics building. The quote goes like this—"As a young man, my fondest dream was to become a geographer. However, while working in the customs office I thought deeply about the matter and concluded it was too difficult a subject. With some reluctance, I then turned to physics as a substitute." The importance and the utility of geography as a subject cannot be explained better than this. As paradigm shift and applications expand, a definition for such a dynamic and all encompassing subject becomes more difficult. But one can readily find the word 'spatial' appearing in all of the definitions in some form or the other. As a subject, geography is concerned with 'where' and 'why' the objects and events are located in space and moreover, with the patterns of phenomena and the processes that created them. In short, geography is the science of spatial analysis or areal interrelations between objects or events as they are observed on the Earth. There is no specific subject matter that it deals with, but rather the subject is our home—the Earth, described and explained with the spatial perspective. Consequently, geography covers a very diverse group of disciplines and their branches in both social as well as physical sciences. At the lower end, it considers objects at the scale of architecture and at the upper end, terrestrial, with a whole range of scales existing between these two.

The major themes in the modern geographical research are identified as (i) spatial association, (ii) spatial forms and processes, (iii) spatial interaction, and (iv) distance decay (i.e., how the magnitude of the variable(s) decreases with distance from their point of origin). There are several key concepts in geography (Dent, 1999) like (a) Direction and true direction (azimuth), (b) Distance between objects (Euclidean system), (c) Scale of the area for study (micro-scale / macro-scale), (d) Location (absolute and relative), (e) Distribution or areal pattern (dispersed/ clustered/ random/ regular), (f) Localization, (g) Functional association (connectivity), (h) Spatial interaction between events and objects with the people, (i) Region, and (j) Change.

As we visualize, geographical phenomena may either be discrete (buildings, trees, roads) or continuous (air temperature, soil pH, elevation). The continuous phenomenon is often described and mapped using discrete units (e.g., land use/ land cover derived from satellite data as a continuous phenomenon is mapped using political boundaries). The dimensions range from zero to four as point (zero-dimension), line (one-dimension), area (two-dimensions), volume (three-dimensions) and space-time (four-dimensions). Based on the origin, there are primary and secondary or derived geographic variables, the latter includes averages (mean, median and mode of primary data), ratios of two variables, proportion (ratio of one variable to the sum of all variables) and percentages of proportions. These variables are measured in terms of qualitative values (*nominal scaling*) like buildings, road, rivers, or on the basis of rank (*ordinal scaling*) like type of drains, road—kaccha / metallic or railway track—broad gauge/ meter gauge. The *interval scaling* adds numerical information to the raw data like rainfall in mm or temperature in degrees. Interval scale usually has arbitrary zero value that does not indicate an absence of variable but rather its value that separates the positive and negative values. There is another measurement scale called *ratio scaling*, similar to interval scaling, but is designated of having a non-arbitrary zero value that indicates a lack of variables in question, e.g., precipitation in mm/year and any point is compared to the starting point (usually zero) unlike to the interval scaling where no starting point is present.

1.2 Definition and Concept of a GIS

With this background let us see what a geographical information system (GIS) is? Attempts have been made to define GIS from various angles. People have generally defined GIS as a powerful computer system with in-built database management component. Thus, GIS is "a powerful set of tools for collecting, storing, retrieving at will, transforming and displaying spatial data from the real world for a particular set of purposes" (Burrough, 1986) or "a database system in which most of the data are spatially indexed, and upon which a set of procedures operated in order to answer queries about spatial entities in the database" (Smith et al, 1987). But to the opinion of some people, a GIS has something more than a computer system. It is a decision support system, which "involves the integration of spatially referenced data in a problem solving environment" or even "an institutional entity" that "integrates technology with database and expertise" (Carter, 1989). Keeping in view of the present day capabilities of GIS and the role it plays, we can define GIS, in our own language, as an information system used to store, organize, retrieve, analyze, output and update georeferenced (or spatially referenced) data, in order to support decision making for planning and management of activities like natural resources and environmental management, transportation and telecommunication utilities, commerce and business affairs, defense services, and various administrative management. Interestingly, the term itself is becoming hybrid and modified to satisfy intellectual, cultural, economic and even political objectives. (DeMers, 2000). Thus here it is Geographic Information System while in Europe it is Geographical Information System. Canadian people call it a Geomatiave (a Geomative). It is Georeltional Information System (technology based) or Spatial Data Analysis System (terminology based) or simply Geoscience (discipline based).

However, there are two basic concepts of a Geographical Information System. One is that the features have attributes associated with them. For example, let us think of a tree. How do we keep track of and communicate information about this tree to other people who need to know all about it? We use a database to keep track of what species it is, how old it is, how tall it is, how healthy it is, and likewise other attributes that are important (like quality and size of fruits). This single tree is one record in a database. We call each category (i.e., tree height) a field. Now let us select a forest where we have many trees and we want to keep track of attributes for all. As we are now dealing with more than one tree, it becomes relevant *where* each tree is located so that we know 'what' information is related to 'which' tree (Fig. 1.1). We map the location of each tree and identify which attributes belong to which tree. Here lies the foundation of GIS. A GIS tells us *where* something is and *what* it is. Geographic location answers to 'where' and attributes to 'what'. Thus any GIS is clearly distinguishing itself from

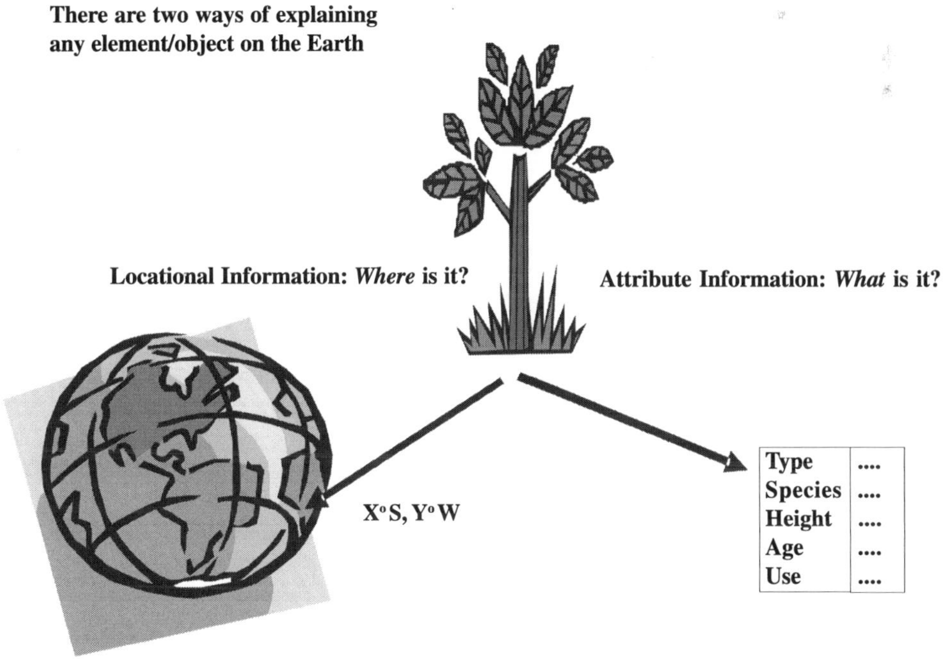

Figure 1.1: Describing our Real World

non-geographic spatial data management systems like computer assisted cartography (CAC) / computer assisted design (CAD) etc., which do not use 'georeferenced' data. As Cowen (1988) has rightly pointed out, CAC systems are excellent for display and can be used specifically for mapping purposes, CAD is excellent for producing architectural drawings and simplifying editing process; but more would be capable of analyzing maps-generally the primary task assigned to any GIS.

The second basic concept is that the information collected must be separated into layers. In GIS, features like rivers, roads or forests are usually stored in different layers so that they can be added or taken off during the GIS project as and when needed. Layers are often called coverages (or databases) and in most cases consist of several computer files that have the same name but different extensions. For examples, in ArcView a coverage is called shape file and can have five files (*.shp, *.sbn, *.dbf, *.sbx, *.prj.) associated with it. Layers represent information of particular classes and can be combined to create new layers containing selected information specific to a particular query on the GIS. Our information on the apple orchard would constitute one layer (or more than one, depending on the objective) of information. We could also have a layer with rivers and a layer with soil types, the purpose may be different. Thus, any information can be represented as a layer. If we look more closely we will identify three fundamental features by which any layer can be represented. These are as follows:

(a) *Points*: A point is good for representing information in which it is necessary to show where a feature is but its physical shape is not important (i.e., location of apple trees in an orchard).

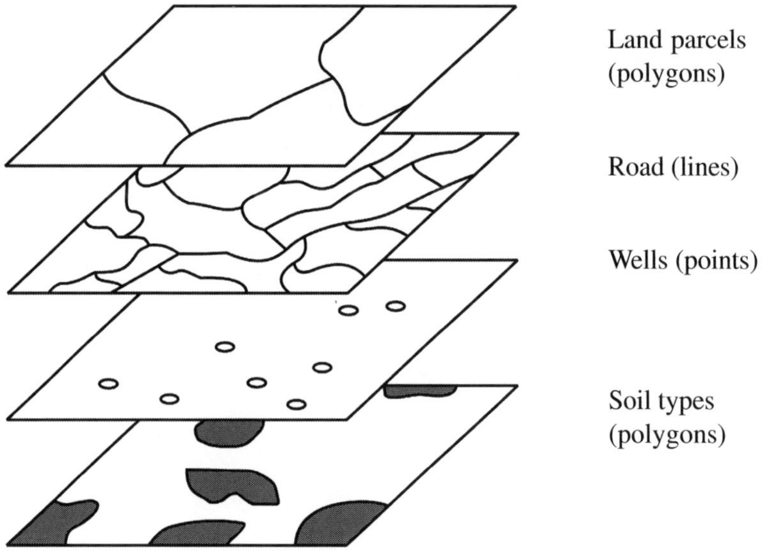

Figure 1.2: GIS layers

(b) *Lines*: A line is suitable to represent any real linear features (i.e., a river or a road).

(c) *Polygons*: It is just a solid multisided shape. When we want to represent an area we have to draw a polygon, keeping in mind that everything inside the boundary has the attributes associated with the record (i.e., land use/ land cover types).

Each layer of information (e.g., location of apple trees, river, land use) has to be tied to the same exact geographic coordinate system so that each layer occupies the exact same space and they can be overlaid. We can think of the layers as floating maps that can be stacked (Fig.1.2) and conduct analysis between layers and display only the layers of interest.

1.3 Components of a GIS

Burrough (1986) suggested that a GIS could have three principal components, namely, computer hardware, application software modules and a proper organizational context. But data being the most important part and the user as the stakeholder, the key components of a GIS should be represented as shown in Figure 1.3.

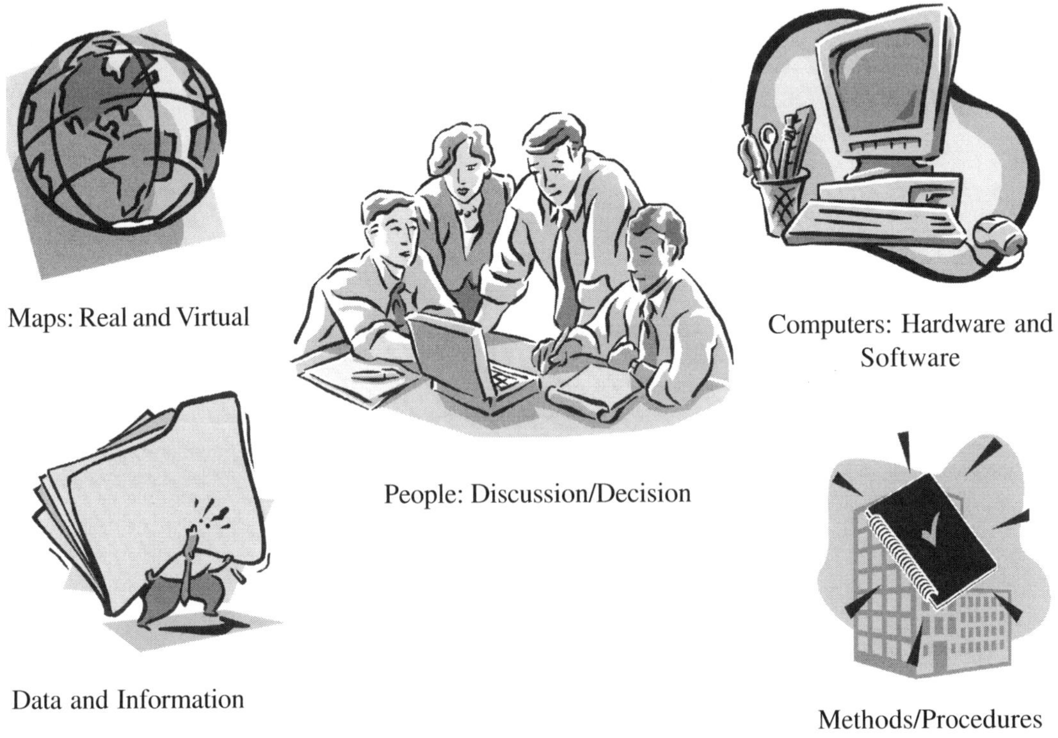

Figure 1.3: Key components of a GIS

The computer hardware should have a powerful processor to run the GIS software, sufficient memory to store large volumes of data and a high-resolution color screen. Apart from these it must have input-output elements like a digitizer tablet/scanner (to convert maps into digital format), keyboard (to enter the data), plotter or printer (as display device) and of course an efficient 'mouse'. GIS software provides the functions and tools needed to store, analyze, and display information about places. The key components of a GIS software are (i) tools for entering and manipulating geographic information such as addresses or political boundaries, (ii) a database management system (DBMS) designed to store, retrieve and update data (iii) tools that create intelligent digital maps for analysis, query for more information, or print for presentation and (iv) an easy-to-use graphical user interface (GUI). GIS software ranges from low-end business-mapping software (usually available as free software which can be downloaded from the internet) appropriate for displaying sales territories to high-end software (commercially available) capable of managing and studying large protected natural areas. The software component must be able to input, verify, store, manage, analyze, output, present and transform data efficiently. Today's software packages are capable of allowing both graphical and descriptive data to be stored in a single database. These packages usually come with a set of tools that can be customized to the users' need. Details of these functions are presented in the subsequent chapters. The type of interface must be user-friendly considering the heterogeneity of the persons handling the GIS. There may be menu-driven commands or typing the commands *via* a command language interpreter. As all operations cannot be menu-driven and the user is required to write their own computer programs, some geographical information systems provide simple and easy-to-understand *macro-languages* that may be used to link many basic applications. The latest example is the simulation modelling with GIS.

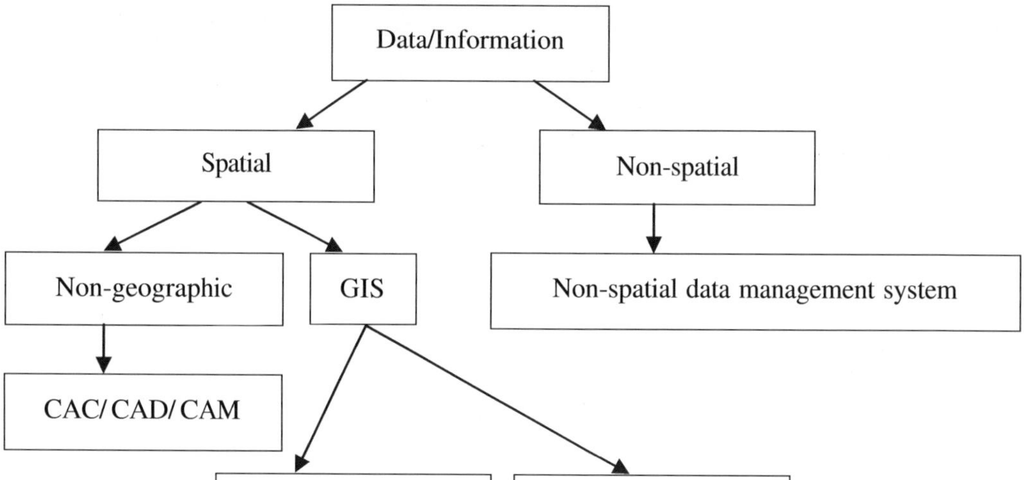

Figure 1.4: Data and GIS system

Data for GIS come in two forms—geographic or spatial data, and attribute or non-spatial data. Spatial data contains an explicit geographic location in the form of a set of coordinates. Attribute data are descriptive sets of data that contain various information relevant to a particular location, e.g., depth, height, sales figures, agricultural production, etc., and can be linked to a particular location by means of an identifier, e.g., address, zip code, district number etc and likewise. Sources of spatial data include paper maps, charts, and drawings scanned or digitized into the system. Coordinate data recorded using a GPS receiver and data captured from satellite imagery or aerial photography are also source of spatial data. Sources of attribute data include database, work flow, messaging and any other form of computer system, which stores data sets that can be linked to the GIS by means of a common identifier. Data streams generated by sensors or data loggers of any kind where the location of the sensor or data logger is stored in the GIS and the data can be linked to it by means of an identifier. Data from satellite imagery derived through image analysis techniques can also be linked to a location or set of locations. Data/information as a component of GIS is explained in Figure 1.4.

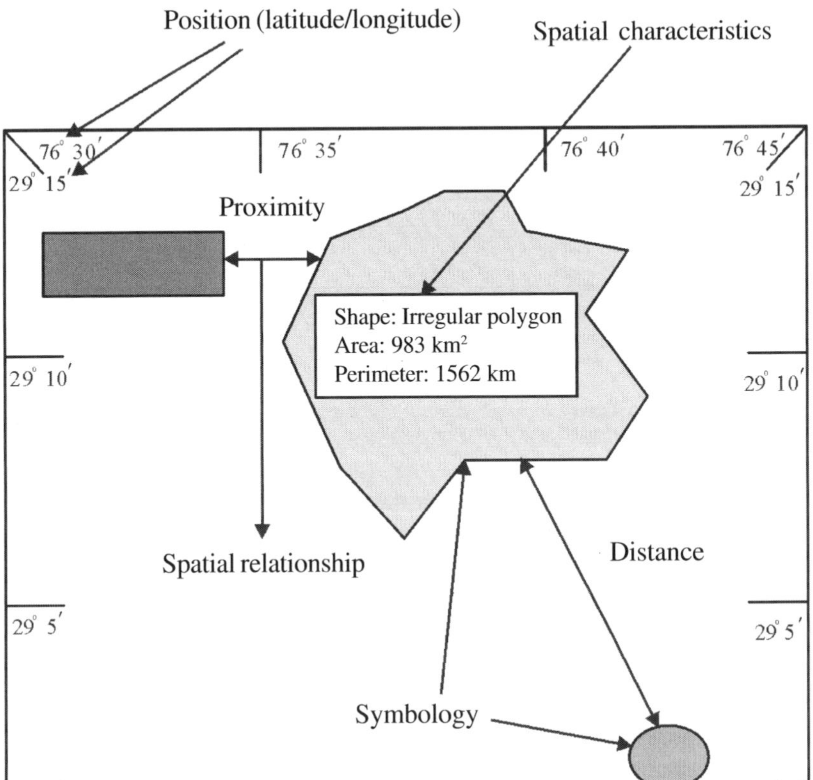

Figure 1.5: Spatial data characteristics

Spatial data has several characteristics (Davis, 1996) like: (a) position: coordinate system on the map that tells the approximate location of an object/feature; (b) spatial relationship: distance from one feature to another (proximity); (c) spatial characteristics: the size and shape of a feature which is defined by the computer with coordinate values; (d) symbology: the line types, point features, and polygon shading interpreted from the legend show the attribute of the features. These characteristics can all be implicitly interpreted, but in GIS these characteristics must be very explicitly delineated (Fig. 1.5).

Data being the most important component of a GIS, the most time consuming and costly aspect of initiating a GIS project is perhaps to create a database. There are several things to consider before acquiring geographic data. The quality of the data must be checked before working on it. Errors in the data set can add many unpleasant and costly hours to implementing GIS and the results and conclusions of the GIS analysis are likely to go wrong. One need to look at least five things regarding good quality data—(1) lineage—the source material from which the data were derived, the methods of derivation, including all transformations involved in producing the final digital files; (2) positional accuracy—this is the closeness of an entity in an appropriate coordinate system to that entity's true position in the system; (3) attribute accuracy—attribute information often includes measurements of some sort, such as temperature or elevation of a point or place. The source of error usually lies within the collection of these facts; (4) logical consistency—it deals with the logical rules of structure and attribute rules for spatial data and describes the compatibility of a datum with other data in a data set. The consistency checks should be run at different stages while handling the spatial data; (5) completeness—this is to see if relevant data is missing with regards to the features and the attributes. This could deal with either omission errors or spatial rules such as minimum width or area that may limit the information.

The fourth essential component is the *user* who plans, implements and operates the system and makes decisions based on the analyzed results. No GIS can exist in isolation from the people and that's why the role of organization is very important in the successful use of GIS in implementing any program. To be useful and successful, GIS must be a part of the organization and must have people and facilities dedicated to it. People also need to be educated to make decisions on what type of system to use. People associated with a GIS can be categorized into viewers, general users, and GIS specialists. The last category of people is responsible for the maintenance of the geographic database and the provision of technical support to the other two classes of users.

The fifth key component of GIS is the methods or procedures. Procedures include how the data will be retrieved, input into the system, stored, managed, transformed, analyzed, and finally presented in a form of output. The procedures are the steps taken to answer the questions need to be resolved. The ability of a GIS to perform spatial analysis and answer these questions is what differentiates this type of system from any other information systems.

GIS is designed and developed to aid the data management and decision support processes of an organization (Fig. 1.6). The operation of any organization is based on a set of practices and business rules unique to that organization. While some organizations may use GIS on an ad-hoc basis with each user formulating their own standards of work and methods of analysis, others define their business logic into the GIS to streamline certain aspects of their operations. For example, an insurance company may integrate a GIS with a workflow system such as Lotus Notes or Microsoft Outlook/Exchange for processing insurance applications. The application will originate in the workflow system by an insurance agent creating a new policy application or a prospective client submitting an application through the web. It will then be passed to the GIS to pick up the appropriate rate using the residential and work addresses given by the client as well as the distances travelled, and other conditions determined by the insurance company in their business logic, the application filled up with all relevant rate and supporting information will be passed back to the workflow system for final approval by an auditor. The integration between a GIS and a workflow system results in significant savings for the insurance company as processes that used to be performed manually by agents are now automated by an integrated GIS solution, resulting in greater efficiency, lower rates of errors, faster response time and a reduction in labor and processing costs, and leading to greater earnings through a reduction in operational costs and improved business due to a higher rate of customer satisfaction. The same methodologies could be applied to any organization that relies on a set of business processes for its operational success, such as banks processing loan applications, retailers managing delivery of goods and supplies, utilities providing repair and maintenance services, government agencies managing permitting and regulatory compliance processes and many others.

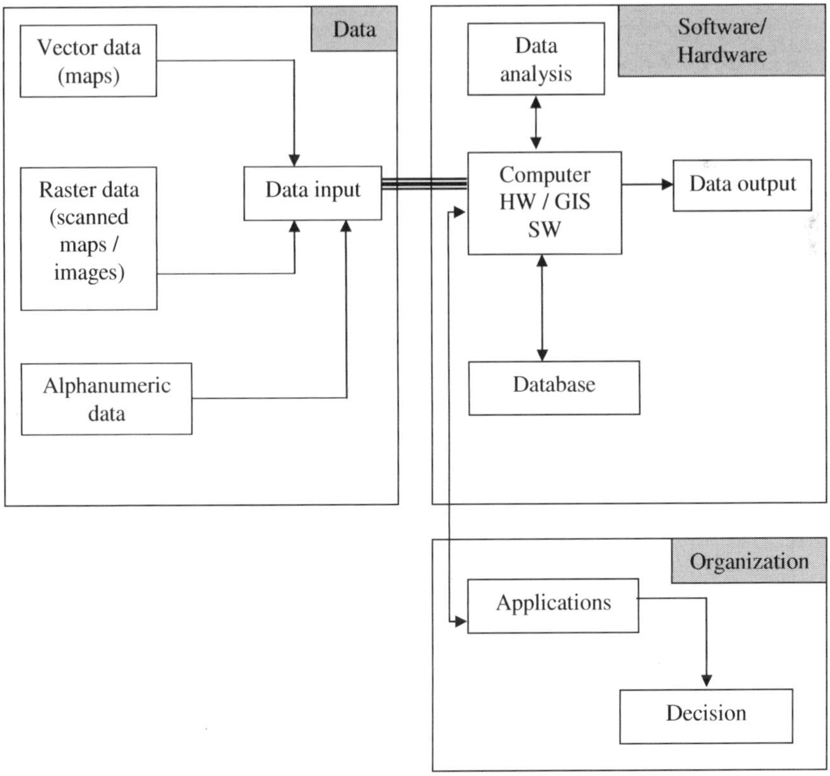

Figure 1.6: Decision making in GIS

1.4 The History of Mapping and GIS

It is not certain about where the future of GIS lies, but now is the best time to ask how it all began (Bill Theon, 1999). In the subsequent chapters we will see how the science and art of making maps evolutionised into a complete science called geographical information science with the advancement of computer systems. Though the history of maps dates back to even before 200 AD, the history of GIS barely spans four decades. In order to have a continuous story of the evolution of GIS, we need to look back to 6th Century BC with the construction of the first Greek map. Box 1.1, provides the time line of history of maps as well as GIS history with special emphasis to India.

	Box 1.1: The Time Line of GIS History
127-148 AD	Ptolemy presented a World Atlas containing about 100 maps.
1531	Gerhard Mercator introduced a globe.
1569	Gerhard Mercator prepared a large World map.
1609	Galileo Galilee gave the concept of the universe with the Earth as its center and all cosmic objects are revolving around the Earth. End of 17th Century: Giovanni Cassini and his son Jacques suggested that the Earth is flattened at the equator.
1687	Sir Isaac Newton confirmed that the Earth is flattened at the poles.
1752	The first detailed map of India was produced by Jean Baptiste Bourguignon d'Anville, a French geographer.
1767	The Survey of India (SOI) was established by the East India Company for mapping its acquired territories.
1781	The Father of Indian Geography, Surveyor General James Rennell published the first ever atlas, the 'Bengal Atlas' followed by the Bihar Atlas and Indian Atlas.
1785	First Map of 'Hindustan' (India) was published.
1802	The Great Trigonometrical Survey of India (GTS) commenced.
1859	The effort by Gaspard Felix Tournachon (also known as Nadar) to make land surveys from aerial photographs was arguably the beginning of remote sensing.
1891	Professor A . Penck put forward the idea of producing a World Map on a scale of 1:1 million at the International Geographical Congress, Berne.
1909	The first aerial motion pictures were taken in Italy.
1910	The International Society of Photogrammetry (ISP) was founded.
1914-1918	The World War-I witnessed a large number of aerial photography.
1922	The International Geographical Union was established in Brussels.
1930	The British Survey of India maps were published on 1:63,360 scales.
1950	Manual map overlay concept described by J.Tyrwhitt.
1956	National Atlas and Thematic Mapping Organization (NATMO) was established in India.

1957 orbiting	With the launch of Sputnik from the then Soviet Union, mounting of cameras on spacecraft became possible.
1958	National Aeronautics and Space Administration (NASA) was established in US.
1959	International Cartographic Association was founded in Bern. Tobler which gave the concept of applying computer in cartography.
1963	Canada Geographic Information System (CGIS) was developed to analyze Canada's national land inventory.
1964	The Harvard Lab for Computer Graphics and Spatial Analysis, Harvard University, US was established by Howard Fisher.
1966	Indian Photo-Interpretation Institute (IPI), presently known as the Indian Institute of Remote Sensing (IIRS), was established at Dehradun. Howard Fisher developed SYMAP (Synagraphic Mapping System), an automated computer mapping application system.
1968	First Geostationary Operational Environment Satellite (GOES) was developed and launched by NASA.
1969	Mettarg presented advanced map overlay concept. Indian Space Research Organization (ISRO) was established at Bangalore. Environmental Systems Research Institute (ESRI) and Intergraph Corporation were founded based on techniques, ideas developed at Harvard Lab and elsewhere.
1972	The first Landsat satellite, ERTS-1 was launched. General Information System for Planning (GISP) was developed by the US Department of the Environment.
1973	Maryland Automatic Geographic Information (MAGI), one of the first statewide GIS projects began in the US. The USGS began development of Geographical Information Retrieval and Analysis System.
1974	The first AUTOCARTO conference was held in Virginia.
1975	India's first indigenous satellite 'Aryabhata' was launched. National Remote Sensing Agency (NRSA) was established in Hyderabad, India for satellite data procurement and distribution. European Space Agency was established.
1976	Minnesota Land Management Information System (MLMIS), another significant state wide GIS, began. National informative centre (NIC) was established at Pune.
1977	The Digital Line Graph (DLG) spatial data format was developed by the USGS.
1978	A Radar Imaging System - the main sensor on Seasat, US was launched.
1979	Indian National Cartographic Association (INCA) was established. The first vector GIS 'ODYSSEY' was developed at the Harvard Lab.
1980	The use of automation and digital cartography started in Naval Hydrographic Office in US (now National Hydrographic Office) with acquisition of new ships, modern automated

	equipments, automated data logging and plotting system, automated cartographic and printing system. ESRI launched ARC / INFO which was the first GIS to take advantage of new super-mini hardware.
1982	Natural Resource Data Management System (NRDMS), a multi-disciplinary program of the Department of Science and Technology (DST) under the Government of India was launched in order to initiate and promote research and development cum demonstration in the field of geographical information systems technology and its applications.
1984	The first International Spatial Data Handling Symposia was held.
1985	The GPS (Global Positioning System) became operational. A raster-based GIS program 'GRASS' (Geographic Resources Analysis Support System), was developed at the US Army Construction Engineering Research Laboratories. Survey of India initiates the Digital Mapping Project to convert 1:50,000 toposheets into digital format for public use.
1986	MapInfo, another GIS package was developed. One of the classic books on GIS, 'Principles of Geographic Information Systems for Land Resources Assessment authored by Peter Burrough was published.
1987	The International Journal of Geographical Information Analysis gets published. SPANS, a raster based GIS was released.
1988	The first Indian Remote Sensing Satellite (IRS) IRS-1A was launched. The National Center for Geographic Information and Analysis (NCGIA) was established in the USA. US Bureau of Census TIGER (Topographically Integrated Geographic Encoding and Referencing) digital data product was released in the public domain. Founded as GIS World, the monthly magazine 'GEO World', the world's first magazine for geographic technology gets published.
1989	The 'Association of Geographic Information' (AGI) formed in the UK.
1991	The first European Remote-Sensing Satellite (ERS-1) was launched that carried radar altimeter. A low cost GIS package, GRAM (Geo-Referenced Area Management)-GIS was designed by Natural Resource Data Management Systems (NRDMS), DST, Government of India.
1993	The European Umbrella Organization for Geographic Information (EUROGI) established in Europe. Indian Society of Geomatics formed at Ahmedabad. National Spatial Data Infrastructure (NSDI) formed in US by an executive order of the US President. PCI Geomatics, a geomatics solution company was formed.
1995	Mapinfo Professional was launched for Windows 95. RADARSAT - SAR satellite was launched.

1996	ESRI-India was formed. Leica India Geo systems Ltd. was established.
1998	CSDMS organized India's first and largest conference and exhibition on GIS/GPS/ Remote Sensing—"Map India 98"
1999	Landsat 7 carrying Enhanced Thematic Mapper Plus (ETM+) was launched . IKONOS launched.

The concept of the Earth as a sphere (and not a flat disc) was first given by the School of Pythagoras. It was brought to general attention through Plato's musings. One can find evidences of definite themes given on maps (like historical incidents) even in those days. A complete section named 'Jambukhandanirmana Parva' in the 'Mahabharata' spoke about the geography (rivers, mountains and 'janapadas') of the then Indian subcontinent. Subsequently, cosmological theories and geographical thinking started evolving. The oldest example of formal cosmography in India seemed to be a chart showing a symmetrical four-continental earth, probably formulated in ancient Brahmanical times. In 127 AD the father of modern trigonometry, Claudius Ptolemy came out with a manuscript that contained topographical details with one world map and subsequently he produced hundreds of regional and local maps during 127 to 148 AD. Since Ptolemy's time till 16th century history witnessed production of maps in different forms and shapes. The most famous of all were the circular world map (mappa mundi) of Hereford, Catalan Atlas, Bianco's and Fra Mauro's world maps to name a few.

In 1531, one of the most learned geographers among all contemporaries, Gerhard Mercator presented a globe, a huge map of Europe and finally a world map in 1569. Mercator introduced an edition of Ptolemy's World Atlas differing completely from the concepts of Ptolemy. Using conformal projection system he recognized and presented three great landmasses, the Old World (Eurasia and Africa), the New Indies (North and South America) and a Great Southern Continent, (Continens Australis) with nearly accurate land surface estimates.

In India land surveying and map-making era started during Akbar's time. These maps basically focused much on the polity of the Mughal Empire like extension of the Mughal territories. The scientific mapping era started at the time of the renowned French cartographer, Jean Baptiste Bourguignon d'Anville (1697-1782) who published a map of India (map of 'Indies') along with maps of North and South America, Africa, Asia, Europe and a two hemispherical maps of the world as a whole. He was followed by James Rennell (Surveyor General of Bengal) and Lt. Col. Mark Wood (of the then Kolkata) among others. The survey of India was established in 1767 by the East India Company to map the territories acquired by the British and it released the first map of India—'Hindustan'. The Great Trigonometric Survey (GTS) of India based on William Lambton's proposal started in 1802 with its aim to produce scientific maps of India to a scale of 16 inches to a mile by the measurement of a base line near Madras. This was followed by a series of triangles across the Peninsula (Great Arc Series). These are the works that had encouraged the Britain, France and Sweden to mathematically compute the exact shape of the Earth.

Box 1.2: Spatial Elements: Point, Line and Polygon

Point data are depicted on images with graphical signs. The size, color, and type of the signs exhibited at point locations may be fixed, or may change according to the fields in a point attribute table. To display this point data one can use circles, squares, diamonds, crosses and other simple symbols. Point features are called discrete because they are able to occupy only a given point in space at any time. Points have no spatial dimension—length or width, even though they may be referenced by their locational coordinates.

Line data are depicted on images with continuous lines, which may be drawn in various styles, thickness and colors. Lines or linear objects are considered as having only one dimension in coordinate space. These objects may be roads, rivers, regional boundaries, fences, or any type of object that is basically long and skinny.

The scale at which these objects are observed places a basic limitation on the ability to visualize them as not having width. Of course, roads, rivers, boundaries, and fences all have two dimensions when viewed at close range. However, the farther and farther away these objects are viewed, the skinnier they become. Ultimately, they will appear so skinny that it is inconceivable to think that they are anything other than linear objects.

Polygons (such as areas) are those objects that engage both length and width. Examples of these two-dimensional objects comprise the area occupied by a hectare, the areal extent of a city, or an area as huge as a continent. In order to describe the positions of areas in space, it is important to realize that they are made up of a series of lines that start and end at the same point. With areas, not only is it possible to denote their position using lines, but also the shapes, orientations, and the amount of territory they occupy can be found. By adding the dimension of height to the area feature allows for the existence of a new form of spatial element. This element is called a surface dimensional object.

The above-mentioned efforts were to produce the general purpose maps or topographical maps, and the cartographers represented real world or administrative boundaries through point, line and area symbols that were very precisely drawn and printed using the existed technology. As the scientific understanding of the Earth and the geographical objects with their interrelationships advanced, different kinds of maps like geology, soils, land cover etc., were on demand. These specific-purpose maps are called as "thematic maps" (though the modern definition of thematic maps includes not only the specific-surface maps, but also maps of single attribute like pH, elevation, air temperature to show their variation over space). These thematic maps had built the foundation of modern-day geographic information science. In as early as 1912, people started using one thematic map to extract data and place it over another. A series of regional maps were prepared by overlaying (not digitally) one map with other (Steinitz et al., 1976).

During 18th Century, the concepts of longitude started evolving. This was started in France, but it was the establishment of the Royal Observatory at Greenwich as a centre for astronomical research to improve navigation techniques, that provided the impetus. Greenwich was being considered as a fixed point (called prime meridian) from which longitude was measured.

Remote sensing, the science of obtaining information of any object or phenomenon without actually being in physical contact with it (Sabins, 1999), has revolutionized the concept of GIS applications, especially in relation to data input and modelling. Remote sensing began its journey with aerial photography. Though Wilbur Wright seems to be the first pilot in remote sensing history who took photographs from an aeroplane, it was Gaspard Felix Tournachon (1859), also known as Nadar, the first person to take aerial photographs of land features from balloons. In 1909, the first aerial motion pictures were taken by Wilbur's passenger, L. P. Bonvillain in Italy. The International Society of Photogrammetry (ISP) was founded in 1912 under the leadership of its first President, Eduard Dolezal from Austria to boost up the research in aerial photography and photogrammetry. The value of aerial photographic survey was only realized during the first World War (1914-1918) when Germany acquired nearly 4000 photos a day during 1918, and the US Army has made over one million prints during the last four months of the war! In 1922, the first layer map of the world was published by the Royal Geographical Society. In the same year the International Geographical Union was established in Brussels. During the Second World War, the importance of well-conceived and conformed maps was again recognized and the science of cartography has since then started developing at a very high speed.

The last 50 years have earmarked a very rapid stride on the development of GIS as a science. Digital map overlay is a new age phenomena, but manual map overlay as a method was first described comprehensively by Jacqueline Tyrwhitt in 1950 in the Town and Country Planning Textbook in Britain (Steinitz et al., 1976). He used four data types (surface geology, hydrology or drainage, farm land area and land elevation) to prepare a single map and termed "land characteristics". In 1962 two scientists in the Massachusetts Institute of Technology (MIT) described the overlay operation by giving suitable weightage to different "themes" based on their relative importance. They generated 26 maps showing the desirability of highways. The procedure was to trace off layer(s) from a selected map or separate type of feature(s) on the map using photographic techniques and combine the layers mechanically. In 1969 Ian McHarg (Professor of Landscape Archeology and Regional Planning, UPenn) used blacked out transparent overlays for site selection, first to introduce modern GIS concepts (McHarg, 1969).

During this period a silent revolution was taking place in the computing world. Ten years before McHarg presented the 'map overlay' concept, W. Tobler, the then a graduate student, put forward the concept of applying computer to cartography (Tobler, 1959). His model was famous as "MIMO" (map-in-map-out) with only three elements, a map input, map manipulation and a map output. Doesn't it sound similar to data capture (data input), data management and analysis and data display as the three fundamental functional units (elements) of modern-day GIS? Though people started writing programs for automatic extraction of maps using printers and plotters none of these programs can actually be called as primitive GIS. As modular computer programming languages (packages with modules linked to one another) developed during late sixties, actual GIS came in. These packages were used to draw chorochromatic (maps showing a set of delineated areas with different color shades) or isoline maps. Overlaying operations became easy and fast.

In India, the automated mapping system and cartographical research started with the establishment of National Atlas and Thematic Mapping Organization (NATMO) in 1956. In the year of its establishment, NATMO produced the first National Atlas of India combining statistical thematic mapping to cartographic knowledge. It also published the first 'National Atlas of India' in Hindi in the following year. During this period National Aeronautics and Space Administration (NASA) was set-up (1958), International Cartographic Association was founded (1959), first meteorological satellite (TIROS-1) was launched (1961) and Canada Geographic Information Systems (CGIS) was developed by Roger Tomlinson for Canadian Land Inventory to analyze Canada's national inventory (1963). Tomlinson was the first person to introduce the term 'GIS'. Some other important systems were Minnesota Land Management System (MLMIS, 1969) and Land Use and Natural Resources Inventory System (LUNR, 1967). But the landmark was set by the establishment of Harvard University's Laboratory for Computer Graphics and Spatial Analysis by Howard Fisher in 1964 and the foundation of Environmental Systems Research Institute (ESRI) in 1969 by Jack and Laura Dangermond as a privately held consulting group. The educational and research program of Harvard Laboratory grew through the 60s, 70s and the 80s with GIS approach and automated mapping systems with development of databases. This laboratory developed many packages important in the developing field of GIS and spatial data analysis namely, Synagraphic Mapping System or SYMAP (a pioneering automated computer mapping application), CALFORM, SYMVU, GRID, POLYVRT and ODYSSEY. Named after Homer's Greek classic 'Odyssey,' the Harvard University ODYSSEY system was influential due to its topological arc-node (vector) data structure. This data structure contains a polygon information with a start node, an end node, an arc between them and the adjacency and connectivity between this polygon and other polygons (Peucker and Chrisman, 1975). Many GIS packages (including Arc/Info) are based on this concept of data model. ESRI's early research and development in cartographic data structure, specialized GIS software tools and creative applications set the stage for today's revolution in digital mapping. Today ESRI continues to set standards in the GIS industry. Its software is installed at more than 100,000 client sites worldwide, of which about 13000 are in Asia and the Pacific regions. The first Geostationary Operational Environmental Satellite (GOES) is developed and launched by National Aeronautics and Space Administration (NASA) in 1968. Indian Space Research Organization (ISRO) was established in 1969. During late sixties and early seventies a space based positioning and navigation system called the Global Positioning System (GPS) was developed by the U.S. Department of Defense (DOD) and started operating very effectively and accurately.

Though incepted for only military use a tragic incident in 1983 in which a Soviet pilot shot down a Korean civilian airliner, lead the Reagan Administration to offer GPS services to the world for the benefit of commercial aviation. The policy of free access to GPS signals was established and became operational in 1985. The NAVSTAR system, operated by the U.S. Department of Defence was the first GPS system to be widely available for civilian use. The Russian GPS system, GLONASS, is similar in operation and is proving complimentary to NAVSTAR system. The European Space Agency (ESA) is now funding the GalileoSat as a new parallel, new GPS setup known as Global Navigation Satellite System (GNSS) to be operational by 2008. This extended GPS system are meant to provide accurate information for critical navigation application.

From late 1960s to till date, large number of geographic data gathering satellites have been launched. Some of the important ones were NOAA (1960), GOES (1965) and LANDSAT (1972). With these satellites in operation, enormous amount of data became available. Tools were needed to analyze patterns (weather, urbanization) or environmental changes (deforestation, ozone depletion, land use, water resources). This has prompted rapid improvement in geographical information sciences in the following years. All these satellite systems are still present with their vast applications for land surface, ocean, weather and atmosphere. National Oceanic and Atmospheric Administration (NOAA) was established at the end of 1970 with many US service and offices like National Environmental Satellite Data and Information Service, National Marine Fisheries service, National Ocean Service etc., that prompted the development of satellite data derived information services in GIS. Series of LANDSAT satellites launched by NASA under the Earth Resource Technology Satellite (ERTS) program made possible to have a unique 25-year data record of the Earth's land surface. Similarly, India made raid progress in space science and especially in the field of satellite data acquisition by sending experimental satellites—Aryabhata (India's first indigenous scientific satellite launched in 1975), Bhaskara, Apple, Rohini and finally series of Indian Remote sensing Satellites (IRS) with remarkably high spatial resolutions. National Remote Sensing Agency (NRSA) is established in 1975 at Hyderabad for acquisition and distribution of data from various satellites (both indigenous and foreign). To keep pace with the development of satellite information, National Informatics Centre (NIC) was set up in 1976 to establish a computer-based informatics network for decision support to various Government Departments for planning and monitoring of development activities. It is also helped to develop database relating to India's socio-economic development through Indian National Communication Satellites (Geostationary) (INSAT) based computer communication network (NICNET) to facilitate information flow from the points of origin to the place of decision making. NICNET is one of the largest VSAT based networks of its kind in the world. Around 700 locations in India, including all state capitals, district headquarters and selected commercial centers can be directly accessed through NICNET. It is connected to over 200 International Networks in 160 countries.

The USGS developed the Digital Line Graph (DLG) spatial data format in 1977. ERDAS was established in the following year. In 1985, after the successful introduction of the National Natural Resource Management System (NNRMS) as a nodal agency for management and monitoring natural resources for their optimal utilization, using space-based remote sensing data in conjunction with the conventional techniques, Indian government with the Department of Space and Ministry of Agriculture set up other two projects namely, National Agricultural Drought Assessment and Management System (NADAMS) and Crop Acreage and Production Estimation (CAPE) for monitoring of vegetation status using NOAA and AVHRR data. In the same year, GPS became fully operational and the first raster based GIS software system GRASS (Geographic Resources Analysis Support System), started at the US Army Construction Engineering Research Laboratories. The International Journal of Geographical Information Analysis got published in 1987 and in the same year raster based GIS software SPANS was released.

In 1988, India started journey in gathering its own remotely sensed data with the launch of its first remote sensing satellite IRS-1A. In 1991-92, many phenomena took place. User-Interface developments led to GIS's vastly improved ease of use during the 1990s. The first European remote sensing satellite (ERS-1) was launched that carried radar altimeter and Japan launched its own satellite JERS-1. In India, Department of Science and Technology under NRDMS designs GRAM (Georeferenced Area Management) GIS for entry, storage, manipulation, analysis and display of spatial data for a low cost computer configuration. This was the first indigenous GIS developed in India. One of the largest Indian private entrepreneurs, TATA started with the Tata Consultancy Services GIS group to provide services in the areas of GIS, Digital Image Processing, Automated Mapping and Facility Management. Integrated Mission for Sustainable Development (IMSD) was initiated, which is an important user of IRS data. Department of Space established Antrix Corporation for overseas marketing of ISRO capabilities in 1992.

In late nineties two very significant achievements in relation to improved data gathering took place. The first one is the launching of Radarsat SAR satellite, an advanced Earth Observation satellite project developed by Canada to monitor environmental change and to support resource sustainability in 1995. SAR (Synthetic Aperture Radar) is a powerful microwave instrument that can transmit and receive signals to "see" through clouds, haze smoke and darkness and obtain high quality images of the Earth in all type of weather at any time. The second remarkable advancement is IKONOS, the first commercial 720 kg high resolution imaging satellite, which was launched into a sun-synchronous, near-polar, circular, low earth orbit on September 24, 1999 from the Vadenberg Air Force base. It simultaneously collects panchromatic and multispectral images of one-meter resolution and is designed to take digital images of the Earth from 680 km up. Moving at a speed of about 7 km per second, the satellite can distinguish objects on the Earth's surface as small as one square meter. The satellite circles the globe 14 times per day, or once in every 98 minutes.

This is not the end of the story. As we finish our journey through the history of present day GIS, it is understood that GIS starting from a mere project, emerged as a system and now evolving into networks (Fig. 1.7). Similarly, scope of GIS software evolved from data to tool to applications, systems and presently to service. The Internet Web drastically increased the amount of data available to the general user and put the GIS on the desktop. Big players like Microsoft, IBM, and Oracle have already entered into the GIS game. With the computing evolution taken place starting

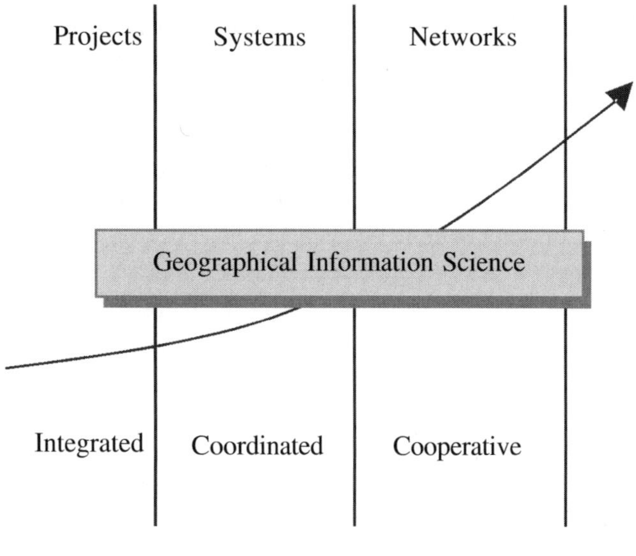

Figure 1.7: The evolution of GIS

with mainframe computers to mini computers, workstation, desktop PCs and finally to internet, GIS software also could more and more exploit the advanced level of computing options. Insidious computing like small (nano) hardware, wireless Internet and interoperable and embedded systems coming into use, capabilities and efficiencies of GIS are likely to increase manifold in the future. Web services will provide a new architecture in GIS integrating all computing systems from desktop PCs/ mobiles to servers. Accordingly, GIS communities are likely to participate in larger networks and thus will evolve from local to global level.

1.5 Why GIS is Needed

Since GIS has arrived now, we need to look back and examine what differences it has brought in today. In a non-GIS world, geospatial data are poorly maintained. Data retrieval, data security and data sharing would have been out of date and inaccurate. GIS makes these things better and easier. Thus, spatial data are better maintained, easy to search, analyze and represent, can be shared and exchanged safely and revision and updating become more comfortable. In this way a considerable amount of time and money is saved, efficiency is improved and at the end we expect a more value-added product and a better decision-making ability. Table 1.1 summarizes some of the benefits in using GIS over manual or conventional system. In addition, some other practical advantages of GIS are listed below.

- Reduction in data redundancy and improvement of response times, thus, development is on time and within the budget;
- Greater efficiencies and economies of scale;
- Improvement of the management of resources;
- Reduction of overhead investment/debt and allowing greater focus on the Core Business;
- Work processes are automated and streamlined;
- More clear visualization and building of relationships between state and federal/ local government, community advocates and agencies and with other community organizations;
- More reliable analysis of trends and documentation;
- Improved tools for Decision Management; and
- Greater sharing of information.

Table 1.1 GIS over manual works

Operations	GIS	Manual System
1. Data input	Encoded into the computer	Recorded or compiled on paper
2. Data storage and digital data base	Fast, integrated and efficient	Slow, very inefficient and different standards for different scales

3.	Data retrieval	Easy, fast and accurate; attribute tables are associated; retrieval requires efficient computer systems	Time and money consuming, inefficient; a method of map reading
4.	Information updating	Computer operated	Manually checking
5.	Map overlay	Systematic	Time consuming and expensive
6.	Spatial analysis	Very fast; completely automated, allows near ocean to raw data and also aggregation and reclassification for further analysis	Time consuming and expensive; Requires tools used by human analyst almost to raw data
7.	Display	Easy, cheap and fast, many options; map is one type of output also there are charts, figures, tables etc.	Expensive, limited options; only map as on output

1.6 Questions that a GIS Answers

Technology can be incorporated into virtually every aspect of life. Its boundaries are essentially limitless. However, there are five basic types of generic questions that any effective GIS can help to answer. Let us review the following questions to see how GIS can apply to any industry or any aspect of life.

A. Identity: What is the … (identity of an entity/feature)?

GIS technology allows users to select or make query of a specific geographic location or feature. For example, users can locate a specific address or property owner. Drawing an analogy—a user can select a specific feature such as a location on a human body and view the related attributes such as organ name, size, function and testing results.

B. Location: Where is… (location of a particular feature)?

A GIS can determine where certain events occur or locate features that have specific attributes. For instance, a GIS can be used to determine which properties are located within a 100 year floodplain. Another example could include locating all the homes of school going children in a given region.

C. Trends: How has the changes occurred (geographic occurrence or trends that have changed or in the process of changing)?

It is often difficult to visualize trends or patterns in data when reviewing tabular data sets. A GIS enables users to graphically visualize their data. For example, police departments utilize GIS technology to track incidences of crime and reveal patterns related to the location, time and types of crimes occurring in the city. In addition, locations of parolees can be displayed in conjunction with crimes to determine if there are any relationships between the two data sets. Other applications often include visualizing the spatial relationships of demographic data sets, such as household income, education and ethnicity.

D. Which data are related (spatial relationship between objects)?

Since features in a GIS are related to each other, it is possible to determine how features are spatially or geographically related to one another. For example, economic development departments may use GIS technology to determine the locations of properties more than 20 acres in size within a quarter mile of a freeway interchange. Other applications may also include locating all properties that are adjacent to easements.

E. What if ... (model based question—to find out the minimum distance, suitable land or risk prone area)?

A GIS can be used to simulate the outcomes of potential scenarios. A large capital improvement project such as a sanitary sewer expansion can be modelled using GIS technology to determine, prior to construction, how the recommended improvements will affect the system's flow and capacity.

1.7 GIS and Other Disciplines

GIS is now emerging as an independent discipline in the name of "Geomatics", "Geoinformatics", or "Geospatial Information Science" that are being used in many government departments, educational institutions and universities. Nevertheless, GIS is said to represent a convergence of technological fields and traditional disciplines like geography, statistics, agriculture, engineering, cartography, operational research, biosciences, remote sensing and so on. It is called an 'enabling technology' because of the potential it offers to the wide variety of disciplines which must deal with spatial data with each field providing some technique that make up GIS. While some fields emphasize data collection, GIS emphasizes data integration, modelling and analysis—leading some to claim it as the science of spatial information. Apart from geography there are several other disciplines, which contribute to GIS today. Figure 1.8 depicts the relationship between GIS and selected four most important disciplines. Table 1.2 summarizes how different disciplines make up GIS with respect to its functions.

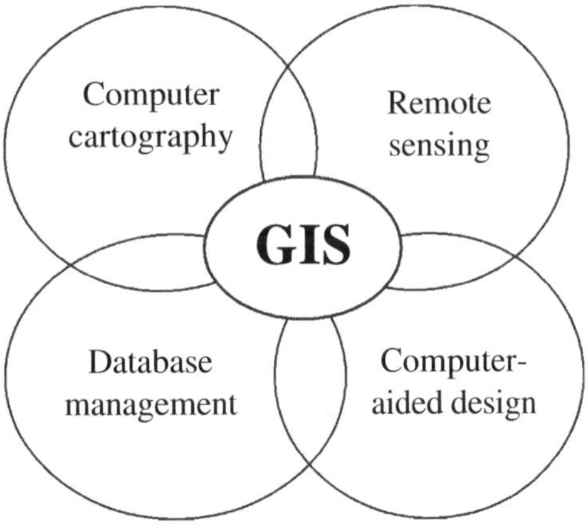

Figure 1.8: Relationship between GIS and some selected disciplines (Source: Maguire 1991).

Table 1.2 Traditional disciplines and GIS

Discipline	GIS functions					
	Data acquisition	Database	Spatial analysis	Modeling	Mapping	Display
Cartography	✓				✓	✓
Agriculture	✓	✓	✓	✓	✓	✓
Engineering	✓	✓	✓	✓		✓
Statistics		✓	✓			✓
Bioscience	✓	✓		✓		✓
Operational research		✓	✓			
Remote sensing	✓				✓	✓
Survey	✓				✓	
Geography			✓		✓	✓

1.8 Areas of GIS Applications

GIS is a wide reaching technology that touches virtually every aspect of life. The GIS industry has grown tremendously over the past 10 years and in effect, it has become an integrated component of our society. One may not even be aware that they are using GIS technology, especially as GIS technology becomes more and more integrated with web technologies. Here are glimpses of some of the most common areas where GIS is being effectively applied.

1. Facilities Management

Large scale and precise maps and network analysis are used mainly for utility management, e.g., network management, locating underground pipes and cables, telecommunication network services, energy use tracking and planning, infrastructure planning etc.

2. Environment and Natural Resources Management

Medium or small scale maps and overlay techniques in combination with aerial photographs and satellite imageries are used for management of natural resources and environmental impact analysis, e.g., natural hazard and disaster management and mitigation, pollution monitoring, environmental impact assessment, waster facility site location, groundwater modelling and contamination tracking, etc., This particular area includes suitability study for crops, management of forests, agricultural lands, water resources, wetlands etc.

3. Street Network

Large or medium scale maps and spatial analysis are used for vehicle routing, locating house and streets, etc.

4. Planning and Engineering

Large or medium scale maps and engineering models are used mainly in civil engineering, e.g., urban and regional planning, rural planning, development of public facilities, route location of highways etc.

5. Land Information System

Large scale cadastre maps or land parcel maps and spatial analysis are used for cadastre administration, taxation, zoning of land use, land acquisition, etc.

6. Financial Services

This includes insurance claim/ risk modeling, demographic profiling, target marketing etc.

7. Health Services

Ambulance and emergency mobilization, epidemiological studies and road traffic accident analysis are some of the services related to health that can be provided successfully and effectively using GIS.

Thus, both the public as well as private sector can apply GIS to its fullest extent. Enterprise GIS allows the public sector to eliminate data silos—segregated pockets of information and results in reduced data redundancy and inefficiencies. By collapsing segregated data into an enterprise GIS that is accessible to many departments, the public sector is increasing the usefulness of the information they collect and maintain. By integrating GIS and web technologies the public sector is also expanding opportunities for intergovernmental cooperation. The Federal or National government, the State government and the Local government can use GIS in their respective areas as follows:

A. Central Government

1. Central Administration
2. Natural Resource Management
3. Defense and Intelligence
4. Internal Security
5. Disaster Management
6. Agriculture
7. Health and Family Welfare Services
8. Housing and Urban Development
9. Transportation
10. Energy
11. Commerce

B. State Government

1. State Administration
2. Agriculture
3. Commerce
4. Health Services
5. Housing and Urban Development
6. Rural Development
7. Environmental Management
8. Natural Resources Conservation
9. Transportation
10. Tourism

C. Local Government

1. Local Administration (like *panchayats*)
2. Local Planning and Area Development
3. Public Works and Utilities
4. Hazard Management
5. Police
6. Housing
7. Recreation

GIS applications in the private sector are also becoming increasingly important. Apart from traditional applications like market analysis, demographic profiling, address matching, sales territory management and site/facility location, companies are now creating web GIS applications which provide the flexibility to easily share information with remote locations, thus reducing the amount of time required to make decisions. The following list of industries is of just a few currently using GIS technologies in their business applications.

1. Archeology
2. Banking
3. Health Care
4. Insurance
5. Real Estate
6. Retail
7. Telecommunications
8. Transportation
9. Utilities

Universities and colleges, libraries and museums use GIS as an important research and analysis tool. Institutional GIS applications spread over a wide range of topics including agricultural, environmental, economic, social and political issues. In addition, as the GIS continues to grow rapidly, more and more schools and universities are offering courses, certificates and degrees in GIS. GIS is even integrated into school curriculum introducing children to geography and problem solving skills through analyzing spatial relationships.

1.9 Critique in GIS

An introduction to GIS as a tool will be incomplete without a discussion of some of its shortcomings and drawbacks. This critique should thus begin with what a GIS is not. A GIS *is not simply a computer system for making maps even though the system produces maps at different scales and projections in a variety of visually attractive colors* (Aangeenburg, 1991). As an analytical tool GIS permits the linking and identification of spatial relationships between features in visual form on a computer screen or a paper map. GIS stores its information in digital form in a series of linked files in a relational database. A GIS is able to retrieve data on particular areas with specific characteristics and show these graphically. In addition, the database management system also links the spatial with other attribute non-graphic, non-spatial data to produce meaningful reports and labels for map features.

So a map that we are familiar with, which previously was stored on our shelves and in our desk drawers now is stored as a digital database on a magnetic medium of a floppy disk or hard disk and which is unreadable and unintelligible to the naked eye as a map. Using the database the GIS will be able to add, subtract, multiply and divide various attributes to produce new relationships previously unseen and perhaps 'unthought' of. The database concept is central to all GIS structures and this distinguishes it from other simple computer aided mapping systems, which may produce good quality maps but can neither highlight the important relationships between features nor offer any form of analysis.

Moreover GIS has the ability to create new relationships, and answer the question, 'what if'—thus functioning as a "decision making tool". However, such a powerful tool when used improperly could produce nonsense results.

Sometimes, GIS is claimed as a new discipline. However, current disciplinary boundaries suggest that this new technology will sit uncomfortably in any one or a combination of disciplines. GIS is predicted to displace cartography, geodesy and other land sciences. The traditional cartographic production process of paper maps is said to be tedious, expensive, slow to produce and costly to update and maintain. GIS itself is in its infancy and suffers the same kind of criticisms leveled against traditional cartography. In geodesy, the lack of architectural scale accuracy even with highly accurate DGPS limits applications at a large-scale. At present even a 1-meter resolution from DGPS is unsuitable for cadastral mapping and land surveys. These very large scale cadastres cannot be used because GIS is ideally suited to small- to medium-scale work. In computer science the rage is with object-oriented database systems where 'objects' are self-contained data and programs which can help in collecting, manipulating and understanding spatial data. As an example, the Windows-based graphical user interface (GUI) uses *icons* (pictures) to depict a file, program or some application. These can be easily 'moved' about the screen using a pointing device such as a mouse. This application, file or data may be used by the user at will by selecting it to start the program. In simple terms these icons are thus said to be objects. The introduction of these ideas into GISs has caused severe anxiety among practitioners and especially among geographers. Many geographical constructs are implicitly uncertain, for example, the concept of 'far away'. Some spatial objects are often the product of interpretation or generalization, for example, proximity, contiguity, randomness, regions and so on. It is therefore very difficult to mould the observations around us into the strict and rigidly bounded objects favored by computer programers and scientists.

A GIS is said to lead to improved spatial analysis. This argument is flawed in two ways. First, there is no way that more data will lead to better analysis. Moreover, in the collection of data there is a large time-lag so that there is always a need to re-evaluate, re-define and adapt to changes recorded and observed. This continual distillation and refinement process is an essential part of the scientific method rather than the static, slice-of-time analysis of the collected data. Secondly, as with traditional methods, GIS can preoccupy the practitioner with the mechanics and technical aspects of the technology that much more time is spent developing the model and tools than the analysis itself. This is a distinct possibility with novice users of GIS. With more expert users the problems with data availability and non-availability of the right kind, in the right form, as well as the quantity and quality of the hardware are some of the ongoing issues to be resolved.

GIS is not easy to use, not only because of the technology but also because of the lack of an industry standard. With so many different vendors of GIS products, there is a lack of a common interface for all users. There is need, on the one hand, to know the command structure of a particular program. For example, the widely used PC Arc/Info has about 2,600 different commands, which a user may need to be aware of if that user wishes to be fully conversant with the system. On the other hand, a user needs to have a broad understanding of the theory and mechanics of spatial analysis. Thus, to be able to use a GIS as a simple 'do-all' tool is a false assumption. There is certainly much more to it than simply producing a map as an end product—it may make a pretty picture but it may not withstand minute and critical scrutiny. The recent advent of a Windows-based version of Arc/Info can provide the much needed standard because all vendors would have to write their programs so that they are compliant with the requirements of Windows programming.

Current GISs are deficient in handling temporal data either to show changes or trends. No models exist to show spatial succession, for example, changing land uses in different locations. The ability to model change through time is constrained by the lack of suitable data and the technology itself to show changes is still unsophisticated.

Organizational issues in the implementation of GIS have seldom been discussed in the literature. This is new and expensive technology, which requires someone in the organization to promote, develop and nurture. Even today, there is limited acceptance and effective use of GIS by government agencies because of the high cost of implementation, cost involved in human resource training and also perpetual fear of data misuse. The latter is a structural problem because people in the higher levels of administration may not wish to abdicate their decision-making roles, while those using GIS at the 'coal-face' may wish to implement certain decisions suggested by the GIS but which do not filer upwards to those who actually make the decisions. These hierarchical decision-making structures may make it difficult for the widespread adoption of GIS. Moreover, there is limited documentation of successful GIS implementation to convince the decision-maker to invest in this new technology.

And some words of caution! Applications of GIS technology are vast, reading out to almost all possible areas. Again the under use of the system capabilities seems to be related more to a lack of understanding of the existing potential of GIS than to the software limitation (DeMers,1991).Thus one must be clear of the task and objectives before actually starting with GIS. Nevertheless, for a person newly introduced to GIS software, the journey has just began. With this lets start our journey through this amazing gift of science and technology and discover its beauties.

CHAPTER 2

Map Language

"A map is the greatest of all epic poems. Its lines and colors show the realization of great dreams."

Gilbert Grosvenor, Editor, *National Geographic*, 1903-1954

2.1 THE LANGUAGE OF MAPS

The literary meaning of 'language' is the communication of thoughts and feelings through a system of arbitrary signals, such as voice sounds, gestures, or written symbols. A *language system* includes some rules for combining its components (such as *words*) and is used by a nation, the people, or other distinct community. The primary purpose of language is to express oneself; to communicate ideas, desires, emotions, etc. Thus, it is more than the use of symbols to represent something in the world. A person who has never got a chance of interaction does not know what a language is. Similarly, he may not make use of the language for effective communication if it is taught as a symbolic language as chimpanzees are taught. We don't grasp a language like a hand grasps an object. We make it an integral part of our daily lives.

The communication through language carries and exchanges information from one person to another, from one place to another. This information may be of different types. It includes text, figures, drawings, table, index, pictures or maps. The map has been a very useful component of communication in our everyday life. It not only provides information in a way to convince people but also acts as a guidance to take decisions. Maps and drawings provide spatial information, to be distinguishable form any other modes of information. Interestingly, we often forget that map is an abstraction of reality, and not a miniature version of reality to show every details. Somehow we think that the map should be a perfect replication and expect to visualize our own hometown in the map depicted on a sheet of paper measuring 25 cm by 40 cm. This type of 'authoritative infallibility' (DeMers, 2000) is perhaps because of map being one of the most common and efficient graphical instrument for communicating information since thousands of years together.

In any database language system the information is stored in a particular way following some rules, for effective retrieval, management and updating of the existing data. Generally, these data does not have any reference to the underlying geography or the spatial relations between objects or phenomena. Thus the population of a country does not say where the country is located (of course if we do not have any prior knowledge of the location) or how far it is from another country. A geographical information system (refer to the definition of GIS in Section 1.2) allows the attribute (non-spatial) information and the spatial component of the map information to be linked together in such a way that we can search the data using either the attribute data (i.e., listing the cities with population density >500 per km^2) or using the map/location information (finding the cities within 50 km of river Ganges). This adds to the searching of information where both the location and the attribute data are necessary like locating the cities within 50 km from river Ganges and with a population density of >500 per km^2.

Since geographical information systems are basically computer systems, they always deal with 'numbers'. Thus we must use any number to specify a location and not as what we do in our everyday life. We refer to locational addresses (which obviously are not numbers) to find a house, or a park or a shop. In GIS, we must use numbers and the most common number description is the latitude and longitude of the location. All locations over a map are described by their respective latitudes and longitudes. It is worthwhile to mention here of another field of science called *cartography*, defined as the science of making charts and maps.

2.2 Cartography as the Root of GIS

Broadly speaking, 'cartography' is the science that deals with the construction, use, and principles behind maps (Clarke, 1999). Though the map-making processes have been in practice since thousands of years, the term "cartography" only came in the mid 19th Century. The traditional approach to mapping, called the 'communication paradigm' (DeMers, 2000) limited the potential of cartography and maps were taken to be the final product only, restricting the communication at this end. On the contrary, the 'analytical or holistic paradigm' (Tobler, 1959) keeps raw data inside the computer and allows to make use of this data to obtain any map-product based on user objectives and news. Thus the computer technology has changed the perspective of maps as an output over the past few decades. Today, maps are both communication and analyzing tools. Regardless of the form of maps as static (printed) or dynamic (virtual) image (see Box 2.1), the aim of maps has always been geographical visualization—seeing earth features in their relationship to space i.e., seeking information in relation to geographical space. Computer cartography has revolutionized traditional cartography in improved map making and visualization of geographic information in a multimedia environment, taking it from the science of geographical visualization to *scientific visualization* (other major components are remote sensing, GIS and GPS). The map-reader travels beyond the printed page and sees the terrain as if he was flying in an aircraft, visualizes the predicted distribution of contamination in the soil profile over time, or models the cropping pattern under the changing climate. Relationships that are difficult to show on a paper map can be easily displayed on a computer screen.

Understanding the way maps are encoded to be used in GIS certainly requires knowledge of cartography. The following sections show how GIS and cartography have strong linkage and why it is often referred that GIS has its roots in cartography.

Box 2.1: Analog Map and Digital Map

A map is defined as "any physical or conceptual depiction of the characteristics that take place on or near the Earth's surface or other celestial bodies" (Campbell). Maps can be of two classes: (1) analog or real maps and (2) digital or virtual maps.

An *analog map* is any tangible map production that has a continuous appearance and may be viewed directly (generally called a hard copy). Traditional drawn or printed productions (ordinarily referred to as maps) fall into this classification, along with maps like aerial photographic productions or the output of other kind of remote sensing productions. Analog maps also include, maps generated employing tools controlled by computers, block diagrams and related drawings, and relief models and globes built to portray some part or all of the surface of the Earth.

Digital maps are associated to analog maps. They have properties that permit them to be transformed into analog maps. Digital maps involve images, which may be viewed directly but are not permanent. An example may be the map image projected onto the screen of a computer monitor. More and more, computer techniques are being applied in mapping and related fields. Computers are being used in the processing and analysis of remote-sensing data, in the construction of maps (known as CAC, computer-assisted cartography), and in the handling of GIS.

Digital map production is able to present several potential benefits, including faster, simpler, and more precise design and construction of maps and a very useful interpretation. Scale and projection changes and the statistical analysis of data, are more readily completed.

2.3 Geometric Models of Earth

Determining the shape of the Earth has long been a difficult problem and a controversial matter. In early days the Earth was described as an *oyster* (the Babylonians before 3000 B.C.), a rectangular box, a circular disk, a cylindrical column, a spherical ball, and a very round pear (Columbus). Ancient Greeks believed the Earth to be spherical, for others it was flat. Eratosthenes (276 BC-194 BC), the great philosopher of Cyrene, North Africa (now Shahhat, Libya) calculated the Earth's circumference and it was the first attempt to produce a World map based on a system of lines of latitude and longitude. The belief that the Earth was flat persisted through ages until Columbus discovered America followed by the voyages of Magellan and the others. Because of surface curvature, mapping the Earth on a flat surface introduced distortion. The mathematician Geradus Mercator devised a map, first used in 1569, which allowed mariners to measure angles directly from the map. But, this mapping stretches the lines of latitude to make the meridians parallel and greatly distorts the land near the poles. Though the Earth can never be flat, its models are still used for surveying over distances short enough so that Earth curvature is insignificant. There may be many more members of the Flat Earth Society than of the American Cartographic Association (Clarke, 1999).

Considering all the history and facts, we still are not able to come to a single conclusion regarding the shape of the Earth. The Earth can be modelled as (i) a *sphere*, (ii) an *oblate ellipsoid* (*spheroid*) or (iii) a *geoid* (Greek word means 'Earth-shaped'). The simplest route might be to assume that the Earth is a perfect sphere. But this is certainly not true, because we can see with our own eyes that the Earth's surface is not sphere - it is covered by hills, valleys, mountains, lakes and so on. Nowadays even the satellite images show Earth as slightly 'pear' shaped and not exactly 'spherical'. Hence, we can think that the Earth's overall shape is 'fairly sphere' like. This term (*fairly*) leads to a very important concept in Geodesy (Geodesy is the branch of Geology that deals with the shape of the Earth and the exact position of geographical points on the Earth surface). We can now break the problem of defining the shape of the Earth into two. One is to define the overall shape of the Earth and the other is to superimpose the mountains, valleys etc., onto the surface of this overall shape.

We use globe to represent the Earth's shape as it preserves the majority of the Earth's shape and illustrate the spatial configuration of continent sized features. (DeMers, 2000). But for all the practical purposes the spherical shape of the Earth must be portrayed in two-dimensional media through techniques collectively called 'map projection' (discussed subsequently in this Chapter).

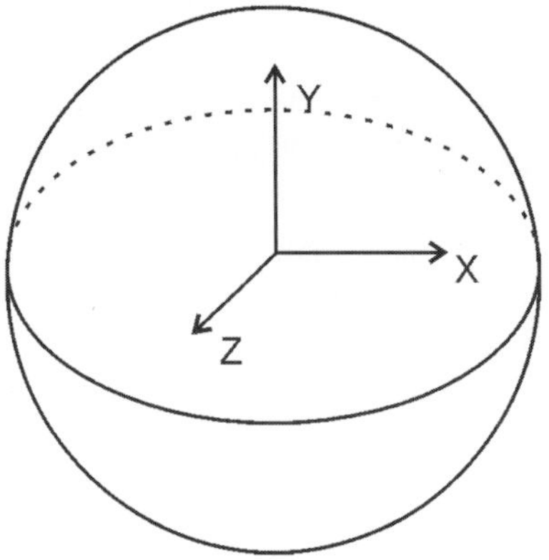

Figure 2.1: A sphere

Spheroids, ellipsoids and geoids all relate to efforts to solve the first part of the problem of defining the overall shape of the Earth. The second part i.e., superimposing the mountains, hills and valleys onto the overall shape of the Earth doesn't really enter into the map projection process. Therefore, we can ignore the second part so long as the map projection is concerned. Now let us assume that the Earth is a perfect sphere. In mathematical terms, a true sphere is defined where all points in three-dimensional space are located at a specific distance (the radius) from the center point. Thus the distances from the center to the surface of a sphere are the same along all the three axes (Rx=Ry=Rz) (Fig 2.1). Hence, in mathematical terms, all one needs to know about a sphere is—location of its center and its radius—and everything is known. This makes map projection processes based on spheres relatively simple. *Spherical Earth Models* represent the Earth as a sphere of a specified radius. These may be used for short-range navigation and for global distance approximations but, as expected, will fail to model the actual shape of the Earth. Since the Earth's overall shape is not very spherelike, so our assumption of the shape of Earth being spherical cannot be realistic. The slight flattening of the Earth at the pols results in about a 20 km difference at the poles between an average

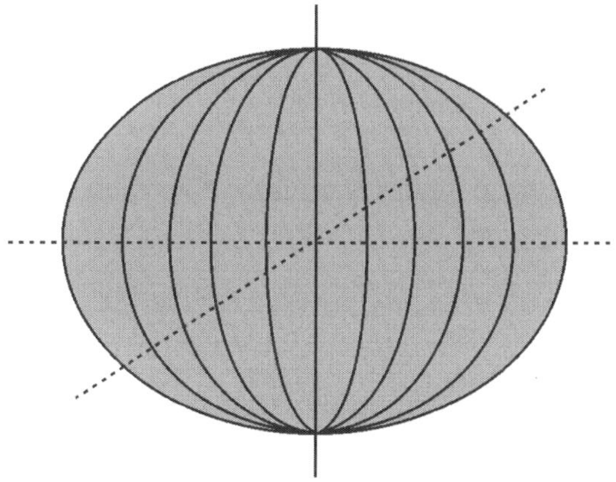

Figure 2.2: A spheroid

spherical radius and the measured polar radius of the Earth. Assuming Earth to be a perfect sphere, the projected map will contain quite a bit of distortion.

In reality, the overall shape of the Earth is more like a *spheroid* than a true sphere. A spheroid (Fig. 2.2) is a sphere like object where only two of the three radii are of equal length (the third radius can be either longer or shorter than the other two). As the Earth is slightly flattened at the poles, the polar axis is shrunk by about 20 km relative to the other two axes. The two equally long radius run through the plane of the equator while the third axis runs through the poles. The spheroidal *models* are required for accurate range and bearing calculations over long distances. The *ellipsoidal models* define an ellipsoid with an equatorial radius (semi-major) and a polar radius (semi-minor) (Fig.2.3). The eccentricity can be computed from the semi-major and flattening parameters. Loran-C, and GPS navigation receivers use ellipsoidal earth models to compute position and to the point information.

Spheroid is actually not the correct mathematical term for this type of *shape.*Technically they

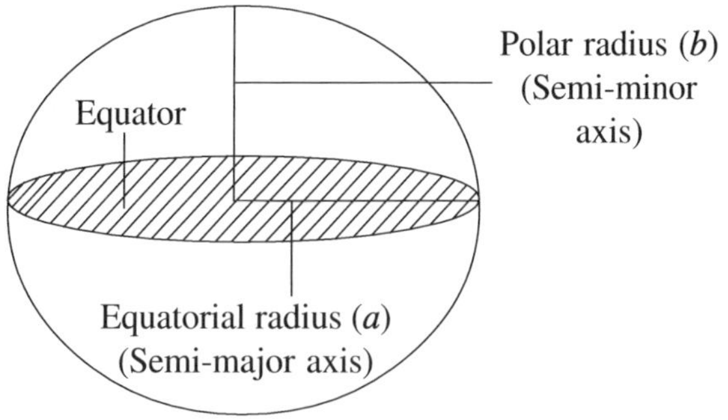

	Equatorial radius (*a*)	**Polar radius (*b*)**	**Flattening ($f = (a - b)/a$)**	**Square of the eccentricity**
World Geodetic System of 1984 (WGS84)	6,378,137.0 m	6,356,752.3 m	1/298.257	~ 0.00669438

Figure 2.3: The Earth as an ellipsoid

are called 'ellipsoids of revolution' (the ellipsoid of revolution is the figure which would be obtained by rotating an ellipse about its shorter axis). Since Earth is flattened slightly at the poles and bulges at the equator, the geometrical figure used in geodesy most nearly approximate the shape of the Earth as an ellipsoid of revolution. In literature, the terms 'spheroid' and 'ellipsoid' are often used interchangeably. Technically this is not correct. For ellipsoid, which is also a sphere like object, all the three axes are of different lengths.

Since the overall shape of the Earth is complex, with lots of irregularities, no spheroid or ellipsoid can provide a completely accurate representation of the overall shape of the Earth. Here comes an important concept in geodesy; namely, the idea of *geoids.*

In principle, the idea of a *geoid* is simple. A *geoid* is nothing more than the true overall shape of the Earth. But in practice, geoids are more complex than what this simple definition suggests. As it is difficult to express or define what we mean by the 'overall' shape of the Earth, so is the precise definition of geoid. The geoid may be precisely defined as a surface along which the gravity potential is equal everywhere (that is why a geoid is also termed as an equipotential surface) and to which the direction of gravity is always perpendicular. Technically this means that a geoid states that at all points, the geoid's surface is located at a distance from the Earth's center of mass where a defined standard object weighs exactly a specified amount. Let us have a brief explanation of the above statement. The weight of any object on earth is determined by the force of gravity (F_G) acting between the object and the Earth. The stronger it is (i.e., the force of gravity), the greater will be the weight of the object. Mathematically, this force is defined by the Newton's law of gravitation (any two objects exert a force of attraction on each other; this force is called gravitational force and the direction of the force is along the line joining the two objects, and the magnitude of the force is proportional to the product of the gravitational mass of the objects, and inversely proportional to the square of the distance between them). Thus force of gravity between any object and the Earth is given by

$$F_G = G\,(M_b \times M_E) / d^2$$

Where, M_b and M_E are mass of the object and the Earth, d is the distance between the centre of the object and the centre of the Earth and G is the universal gravitational constant.

$$G = 6.67 \times 10^{-11} \text{ Nm}^2 / \text{kg}^2$$

As expected, when d increases, the value of F_G decreases. In other words, as the distance between the centre of the Earth and a body/object increases, the force of gravity diminishes. Thus, if an object is placed few thousand kilometers away from the center of mass of the Earth, the overall force of gravity between the Earth and the object would be very small, and the object would weigh very less. Conversely, if the object comes close to the Earth's center of mass, the overall force of gravity between the object and the Earth would become very high, and the object would have a very large weight. Thus, the weight of a body is relative to its position (distances) from the centre of mass of the Earth.

We see that the geoid is defined in such a way that the object weighs exactly the same at all points on the geoid's surface. If the Earth's mass was uniformly distributed throughout the planet, the geoid would be a perfect sphere. However, since the Earth's mass is not evenly distributed, the surface of the geoid undulates. This uneven distribution of mass gives the geoid its complex shape (Fig. 2.4).

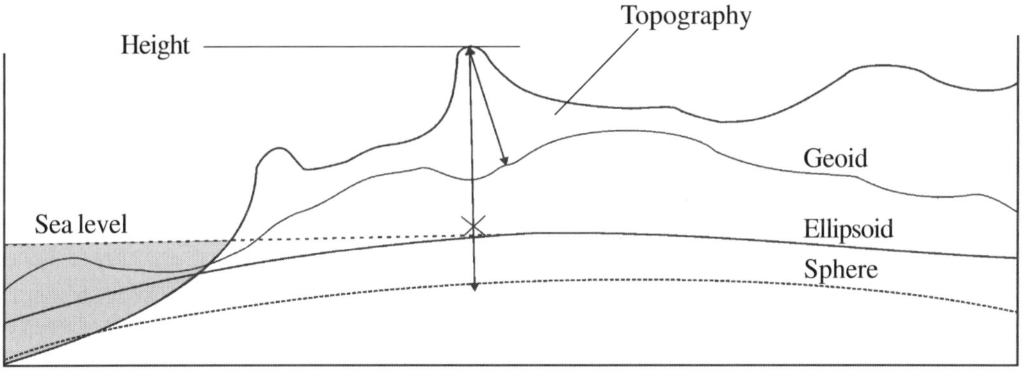

Figure 2.4: Representation of the Earth surface

Again, according to the definition, the direction of gravity is always perpendicular to the geoid surface. This is particularly significant because optical instruments containing leveling devices are commonly used to make geodetic measurements. When properly adjusted, the vertical axis of the instrument coincides with the direction of gravity and is, therefore, perpendicular to the geoid. The angle between the plumb line which is perpendicular to the geoid (sometimes called "the vertical") and the perpendicular to the ellipsoid (sometimes called "the normal") is defined as the deflection of the vertical.

Geoid is an extremely complex shape. The U.S. Defense Mapping Agency tried to develop an exact mathematical model of the geoid using modern satellite and statistical techniques. Even then the model is very complex with 32,755 coefficients and terms ranging up to the 180th power. But, nevertheless, geoid is the only model that is the nearest to the actual, true overall shape of the Earth.

2.4 THE EARTH SURFACE

Based on our previous discussion it is imperative that we need to concentrate on three shapes: spheroid, geoid and the actual shape of the Earth. Spheroid is not a perfect model to present the overall shape of the Earth, but we know its dimensions exactly and hence can use it in the projection process. Then we have the geoid—the shape is very near to the actual, used in the projection process, but is incredibly complex and difficult, as we do not know its true dimensions. We also have the shape of the true surface of the Earth (topography or terrain), which is again infinitely complex and impossible to model exactly. As clearly indicated in Figure 2.4, elevations defined with reference to a sphere, ellipsoid, geoid, or local sea level will be different. Even location as a given latitude and longitude (defined later) may also vary. Here it becomes important to know the base to be used when linking field data such as GPS (Global Positioning System) with a GIS. Errors are more at large-scale (encountering finer details) as the elevation differences would be large at these scales.

Models of the surface of the Earth are used in navigation, surveying, and mapping. Topographic models attempt to mimic the physical variations of the surface, while geoids are used to represent local variations in gravity that changes the local definition of a level surface. The topographical surface of the Earth is the actual surface of the land and sea at some moment in time. This may change over time. Sea level (or the mean sea level, msl) is the average surface of the oceans at a particular time and in relation to a specific method of measurement. Tidal forces and gravity differences from location to location cause this sea surface (considered very smooth) to vary even by hundreds of meters across the globe. Details of the variations in the gravity field are related to the idea of leveling. Plane and geodetic surveys uses the idea of a plane perpendicular to the gravity surface of the Earth, the direction perpendicular to a plumb bob pointing toward the center of mass of the Earth. Local variations in gravity, caused by variations in the Earth's core and surface materials, cause this gravity surface to be irregular. Geoid models attempt to represent the surface of the entire Earth over both land and ocean as though the surface resulted from gravity alone. The resulted surface considering all local variations from the 'spheroid' is a 'geoid' and this geoid has been mapped already. For example, the World Geodetic System 1984 (WGS-84) defines geoid heights for the entire Earth and is accepted by the U S Military Service along with the Geodetic Reference System (GRS80).

2.5 Coordinate Systems and Reference

Knowing "location" is the first step towards understanding the spatial concept. Locating an object needs a complete structured mechanism to follow in relation to a particular reference system (s). Once the reference system is fixed, the absolute or true location of any point in space can be defined by structurised values known as *coordinates*. An understanding of the reference coordinate systems is more critical when a GIS user builds the databases from many different sources (like conventional maps, sattelite imagery, digital date sources, GPS etc.). Coordinates have been used for centuries, both in the western and eastern geodesy to identify locations on the surface of the Earth. Today there are many different coordinate systems, based on a variety of geodetic datum (explained later), units, projections and reference systems. The need for use of more than one coordinate system arises from the fact that many different physical phenomena can be calculated or understood easily in a system that may be appropriate for the phenomenon. Thus, there is often a need to transform one coordinate system to another. Basic coordinate systems can be broadly grouped into Plane Coordinate Systems (Cartesian or Polar) or Global Coordinate Systems. These are briefly discussed here with suitable examples.

2.5.1 Plane coordinate systems

To define a coordinate system in three-dimensional space, we have to specify the origin and the direction of one of the axes, as the orientation of second axis will be in the plane perpendicular to this direction. The third axis follows automatically in order to complete a right-handed orthogonal set. René Descartes (1596-1650) introduced this system, often referred to as Cartesian system. Similar systems based on angles from baselines are often referred to as polar systems. Depending on the dimensions, the coordinate systems can be either plane (single plane, two-dimensional coordinate systems) or three-dimensional coordinate systems (two orthogonal planes, three-dimensional) (Fig. 2.5a and b).

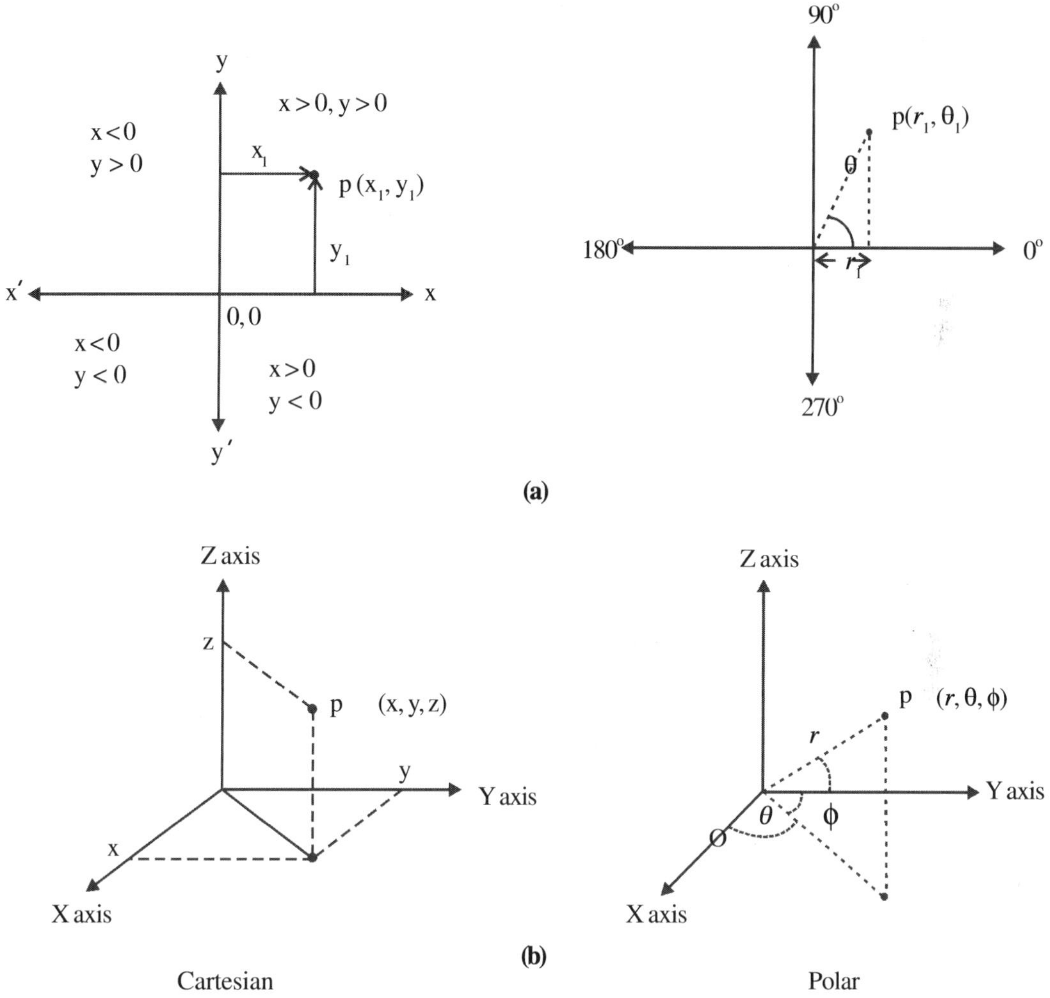

Figure 2.5: (a) Single plane coordinate system; (b) Three-dimensional coordinate system

2.5.1.1 Plane cartesian coordinate systems

As stated earlier, Cartesian coordinates (or rectangular coordinates) are determined by locating an origin and then setting two axes through origin in fixed directions, at right angles to each other. Cartesian coordinates are usually referred as x and y by convention (x is horizontal, y is vertical and anticlockwise from x) as given in Figure 2.5. The linear displacement of any point from the origin in directions defined by the two axes is measured and the location of this point is defined by an ordered pair (x, y). The point x = 0, y = 0 is the ***origin***, where the axes intersect. Cartesian coordinates can be used directly to calculate distance between two points. For example, Euclidean distance (*D*) from point (x_1, y_1) to point (x_2, y_2) will be

$$D = (x_1 - x_2)^2 + (y_1 - y_2)^2$$

provided there are no barriers on the route.

One must take care of the size of the study area and the resolution at which the measurements are taken while using Cartesian coordinates to define the spatial locations of objects. If the size of the study area is large and the resolution is high, more decimal or binary digits to store the coordinate values are necessary. The distance between two points in reality may not be as precise as their coordinate values. For example, for an area as the size of India, to specify locations precisely to the nearest meter, we may require 5-6 digits, i.e., the coordinates of two points may be x_1 = 12050, y_1 = 33200 and x_2 = 12050, y_2 = 33250. So for a distance (in the y direction) of 50 m, the coordinates must be stored with 5 digits of precision, making the accuracy of the coordinates as 1 part in 100 000, but the accuracy of the distance being 1 part in 100. However, to store the distance as 5 digits of precision (as floating number), there will be unnecessary memory usage. In the above example, distance is more uncertain and inaccurate as compared to the coordinates. Thereby it is absolutely necessary to keep track of how results, like distances, are affected by database accuracy.

2.5.1.2 Plane polar coordinate systems

In polar coordinate system the location of a point is characterized by two numbers (Figure 2.5): the distance from origin or pole (0,0) *r*, and the angle θ from fixed direction originated from (0,0) (usually towards north and the angle is measured clockwise from the direction). The angle θ is a multiple of 360°. Sometimes, when *r* is greater than zero, (*r*, θ) and (–*r*, θ + 180°) represent the same point.

The Cartesian and polar coordinate systems can easily be transformed from one to the other using the relation as:

$$x = r \cos \theta, \quad y = r \sin \theta$$

$$r = (x_2 + y_2)^{1/2} \quad \text{and} \quad \theta = \arctan (y/x)$$

2.5.2 Global coordinate systems

Considering Earth as a spheroid, a spherical grid system can be created over its surface using simple geometry (Fig. 2.6) to obtain two sets of imaginary lines. One set of lines are parallel to each other (named parallels) encircling the globe from east to west. The greatest (in length/circumference) of them is at the middle of the Earth (named *Equator*) and assigned a value of zero, the other set of lines (not parallel to each other but parallel to the *parallels*) run from north to south poles and are called the *meridians*. The zero point of the meridians runs through Greenwich, England and is called the *Prime meridian* and also the 'International Date Line'. The prime meridian and the equator are the reference planes used to define the latitude and the longitude (Fig. 2.7). The latitude of a point is the angle from the equatorial plane to the vertical direction of a line at this point normal to the reference ellipsoid. The longitude of a point is the angle between a reference plane at the prime meridian and a plane passing through the point, both planes being perpendicular to the equatorial plane.The height at a point is the distance from the reference ellipsoid to the point in a direction normal to the ellipsoid. These three parameters, latitude, longitude and altitude (height) constist global coordinate system.

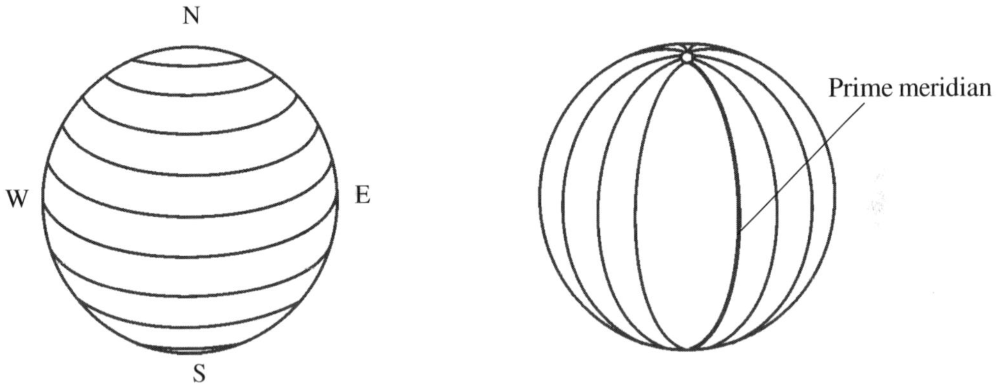

Figure 2.6: Lines of latitude (parallels)

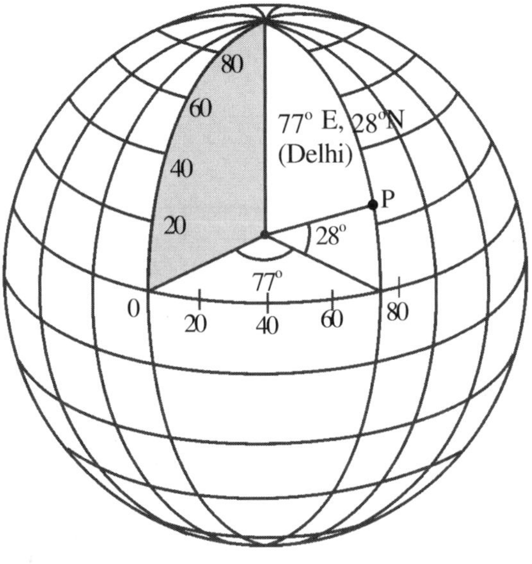

Figure 2.7: Latitude and longitude of a point (Delhi) on the Earth

Latitude and longitude are necessary to locate objects on the sphere (Earth). But as we mostly use two-dimensional maps projected from the spherical Earth or reference globe, other coordinate systems are necessarily required to counter the distortions occurred in the process of projection. Hence plane coordinate systems are more useful to locate objects on the flat maps more correctly than global or geographic coordinate systems as the former use linear values, and not angular values. Here is a word of caution—plane coordinate systems provide clear results for small-scale maps, unless adjustments are made to correct the distortions during projection (Robinson et al., 1995; DeMers, 2002). Some of the important plane coordinate or grid systems are briefly discussed below.

2.5.2.1 Universal transverse mercator (UTM)

Universal Transverse Mercator (UTM) is the most common coordinate system being used today for any database development in GIS. UTM provides very precise measurement using the metric system (meter as the unit measurement). Developed from Mercator cylindrical projection in 1977 by Johann Heinrich Lambert and later on modified by Gauss (1822) and Kruger (1912, 1919). This system was adopted by US Army in 1947 for designating rectangular coordinates on large scale military maps. This system is now adopted by many countries for a wide range of applications. This system divides the Earth from 84° N latitude to 80° S latitude into 60 vertical zones with each longitudinal strip of 6 degrees wide. The UTM has two primary starting points, one at Equator (0° North) and another 10,000,000 m south of the equator. These zones are numbered 1 to 60, starting at the International date line (180° longitude), and proceeding towards

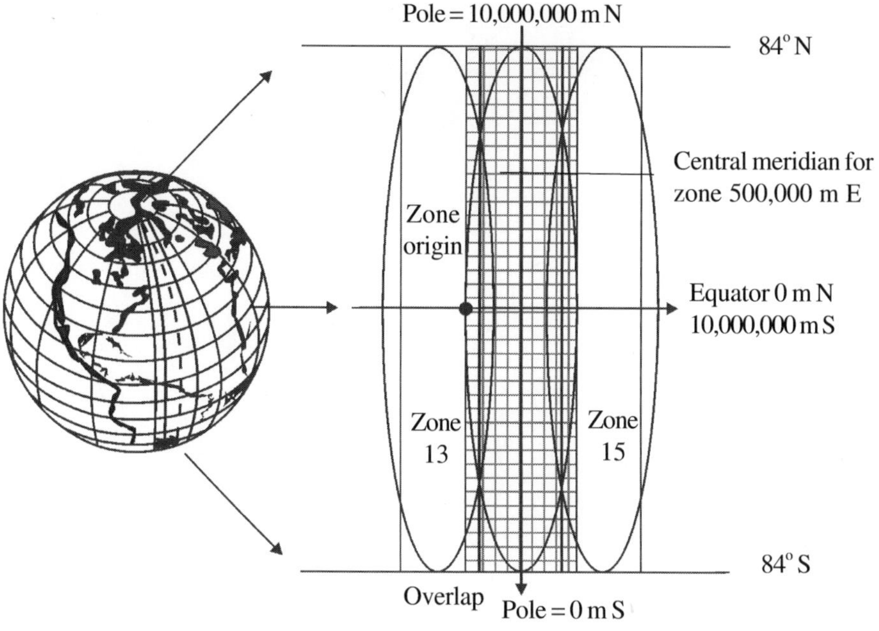

Figure 2.8: Universal Transverse Mercator system

east (e.g., Zone 1 extends from 180° W to 174° W and is centered on 177° W).

Each UTM zone is again divided into rows or bands of 8 degrees of latitude (except the north most section which is 12 degree). These horizontal bands are lettered (Fig. 2.9), south to north, beginning at 80° S with the letter C and ending with the letter X at 84° N. The letters I and O are skipped to avoid confusion with the numbers one and zero. The location of a point in a zone is specified as easting (x coordinate) and northing (y coordinate) and expressed in meters. The central meridian (in each tone) is assigned a false easting value of 500,000 m, only to avoid negative numbers at the western edge of the zone.

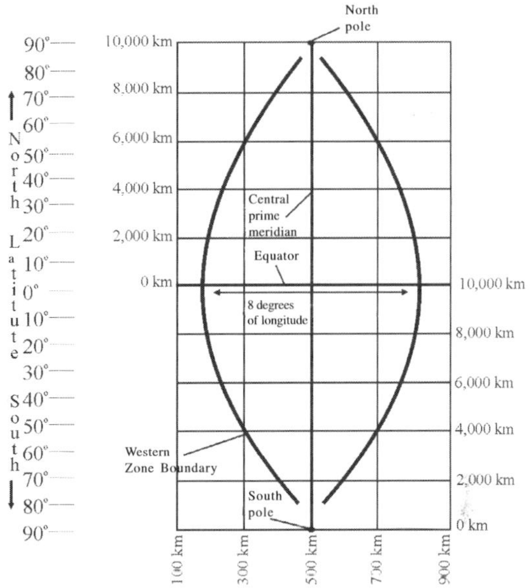

Figure 2.9: A single zone of UTM

The popularity of UTM coordinate system is mainly because:

1. It is a square grid. Unlike to the lat-long coordinate system, the distance does not change when one moves toward poles.
2. Grid values increase from left to right and bottom to top and there is no N-S or E-W directions.
3. Only a few zones can represent a large area (e.g., India occupies only 6 UTM zones).
4. No conversion from degrees to minutes and to seconds. All coordinates are decimal based.
5. Being a conformal projection, this grid system allows minimum directional errors in a zone.
6. UTM is expressed in metric system, which has been universally adopted.
7. One of the very popular geodetic measurement system, GPS (Global Positioning System) provides the coordinate measurement in UTM; similar is the case with most of the GIS software today.

As the name suggests, the UTH grid system uses a transverse Mercator projection for each and every vertical zones repeatedly to avoid distortion. The Y-coordinate origin is placed at the exact centre of each zone with a false origin at 3° west. Scale errors covered by the mercator projections do not exceed the tolerence value and the distortion with UTM zones vary minimum (1 m in every 25000 m).

2.5.2.2 Military grid reference system (MGRS)

The Military Grid Reference System (MGRS) is an extension of the UTM system (Fig. 2.10). Unlike UTM which has three parts, zone, easting and northing) MGRS has one additional part, a *latitude band*. A latitude band is a single letter from C through X depending on the latitude. Sometimes this letter and the *Zone* are regarded as a single entity, the *Grid Zone Designation* or GZD, with the latitude band specified by the *Zone designation letter* or ZDL. The world is generally divided into 6° (east-west) by 8° (north-south) geographic areas, each of which is given a unique identification, called GZD. These areas are covered by a pattern of 100,000-meter squares. An *MGRS square* is defined by two letters, and representing a 100 km square within the zone. Starting at 180 degree meridian the characters A to Z (not I and O) are used for 18° before starting over. From the Equator north the characters A to V (again not I and O) are used for 100 km squares, repeating every 2,000 km. An *easting* measuring the distance from the western edge of the MGRS square to the point of interest and a *northing* measuring the distance from the southern edge of the MGRS square to the point of interest.

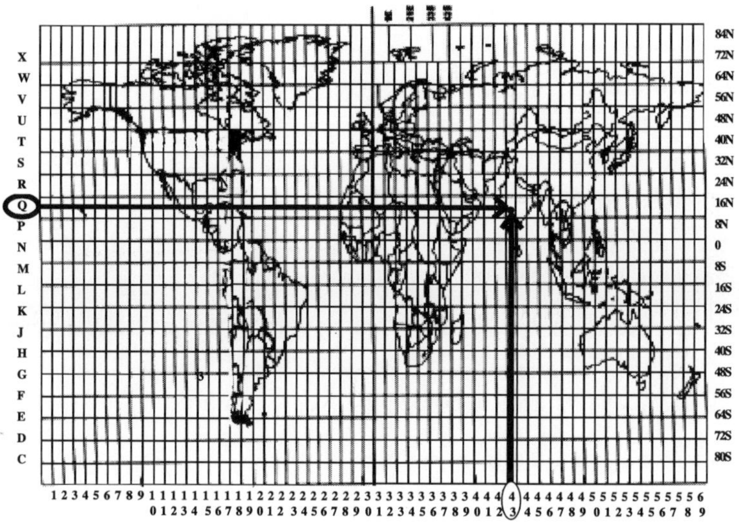

Figure 2.10: Six by eight-degree cells on the UTM military grid

2.5.2.3 World geographic reference system (GEOREF)

The World Geographic Reference System, a simple, rapid, worldwide position reference system is used for aircraft navigation and air defense system. GEOREF may be applied to any map or chart with latitude and longitude, regardless of its projection. GEOREF coordinates are divided into three divisions. The first division of GEOREF divides the surface of the Earth into 24 longitudinal (north-south) zones of 15° each and 12 latitudinal bands of 15° each. These longitudinal zones, starting at 180° and progressing eastward, are designated from A to Z (here also I and O omitted). The latitudinal bands are lettered from A to M (omitting I) northward from the south pole

(Fig. 2.11). The zones and bands divide the Earth's surface into 288 quadrangles, each identified by two letters. The first letter of any GEOREF coordinate identifies the north-south zone (longitude) and the second letter identifies the east-west band (latitude) in which the point is located.

In the second division, the 15° quadrangle (first division) is subdivided into 225 quadrangles formed by 15 longitudinal (north-south) zones and 15 latitudinal (east-west) bands each with 1° wide. The north-south zones are lettered A through Q (omitting I and O) from west to east. Similar is the case for the east-west bands. The third and fourth letter of any GEOREF coordinate identifies 1° north-south and 1° east-west zone, respectively, within a 15° quadrangle. For third division, the 1° second division quadrangle is again divided into 60 longitudinal zones of 1° east and west (numbered 00 through 59 from west to east); and 60 latitudinal bands of 1° north or south (numbered 00 to 59 from south to north) to make 3600 third division one-minute quadrangles. Thus, any geographical point on the Earth's surface can be located into within an accuracy of 1 minute by using a four-letter and four-digit grid reference.

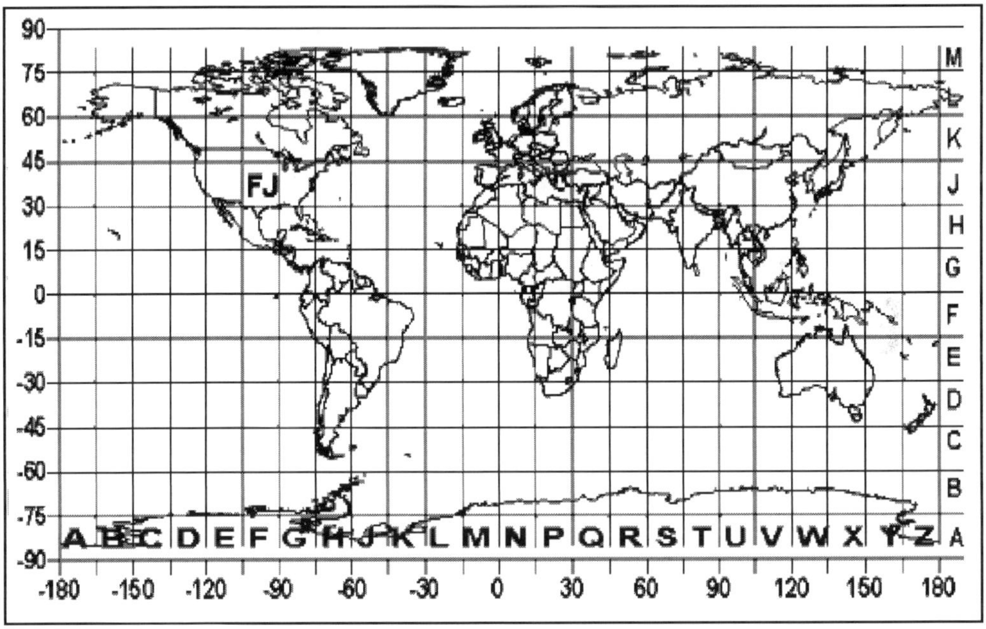

Figure 2.11: World geographic reference system (GEOREF)

2.5.3 Regional/Local Systems

Apart from the global coordinate systems, several different systems are also used locally or over a specific region. Some of these are true coordinate systems (based on UTM and UPS) like British National Grid (BNG) while others are very local and used for specific purpose like US Public Land Survey system (PLSS) or Coordinate Geometry (COGO).

2.5.3.1 Universal polar stereographic (UPS)

Any location in the polar regions (above 84° N and below 80° S latitude) are usually referred by Universal Polar Stereographic (UPS) grid system. This area is beyond the purview of UTM system. UPS system divides polar regions into concentric zones and then splits them into two equal halves (east to west at longitude 0 and 180°) with different designations. In the northern hemisphere, the west half is designated Y and east half zones, while in the southern hemisphere these are named A and B, respectively. Measurement are made in easting and northings, similar to UTM system.

There are some National Grid Systems for different countries based on coordinates that cover their territory. Australia, Belgium, England, Ireland, Italy, Netherlands, New Zealand, and Sweden are the examples of nations that have defined their own National Grid System. There are also State Plane Coordinate Systems like that existing in the US which are designed for engineering purposes. For India, Lambert Grid system is adopted which is again based on conformal projection, Eight lambert grids cover India and Burma (named as I, IIA, IIB, IIIA, IIIB, IVA, IVB, and O). The N-E extent of the grid is 8° while the E-W extent is limited to 16°. The grid origins are spaced 7° apart in latitude with 1° overlap at borders.

2.6 Geodetic Datum

For maps covering very large areas, the Earth may be assumed as perfectly spherical. Conversely for very small areas, terrain features dominate and measurements cannot be based on a flat Earth. For highly precise maps of smaller regions, accurate measurements are necessary which requires some reliable starting point which is referred to as a datum (DeMers, 2001).

A datum involves detailed survey of a country, continent, or some other portion of the Earth's surface. As a part of the survey, the longitudes and latitudes of a large number of points are measured using the best available methods. Once the survey is complete, maps of the region can be constructed. Most regions of the Earth, as on today have been surveyed many times, possibly with increasing accuracy. Each of these surveys is referred to as a datum.

The first estimates of the Earth's size were made by Aristotle. Since then datums have evolved from those describing a spherical-Earth to ellipsoidal models derived from years of satellite measurements. Modern geodetic datums range from flat-Earth models used for plane surveying to complex models used for international applications which completely describe the size, shape, orientation, gravity field, and angular velocity of the Earth. A common datum allows the geodatic scientists and surveyors to compare their measurements among themselves. While cartography, surveying, navigation, and astronomy all make use of geodetic datums, the science of geodesy is the central concept. Referencing geodetic coordinates to the wrong datum may result in positional errors of several hundreds of meters. Different nations and agencies use different datums as the basis for coordinate systems used to identify positions in geographic information systems, precise positioning systems, and navigation systems. Hence global positioning measurements with sub-meter accuracies requires careful datum selection and conversion between the coordinates in different datums.

As the accuracy of measurements involves overtime, so do the datum. It is, thus not possible to replace a completely old datum with an new overnight. Therefore, at any given time and for a given region, more than one datum are in use as introductions of a new datum involves time, money and labor. For example, in the United States, both the North American Datum of 1927 (NAD-27) and the World Geodetic System of 1984 (WGS-84) are in widespread use. In addition to these, maps and data in numerous other datums are also available.

The most recent, and most accurate datum is the World Geodetic System of 1984 (WGS-84). It is a modified version of GRS 80 (geodetic reference system) which is again synthesis of a new datum, NAD 83, established in 1983. While improvements in surveying techniques in future may result in modifications to this datum, these are expected to be minor and therefore, for most practical applications, WGS-84 is likely to remain the ultimate datum.

If errors of several hundred meters are acceptable, one can ignore the entire issue of datums. However, if one requires greater accuracy, he must know the datum of any geographic data being used. Whenever geographic data (maps, computer files, tables of locations) are obtained one has to be sure that the datum of that data is required. For maps, the datum can often be found printed in the legend area of the map. For computer files, the datum is often listed in the accompanying documentation. However, if the datum of somebody's data can not at all be determined, the WGS-84 should be assumed. However, there would be potential errors if this assumption is wrong.

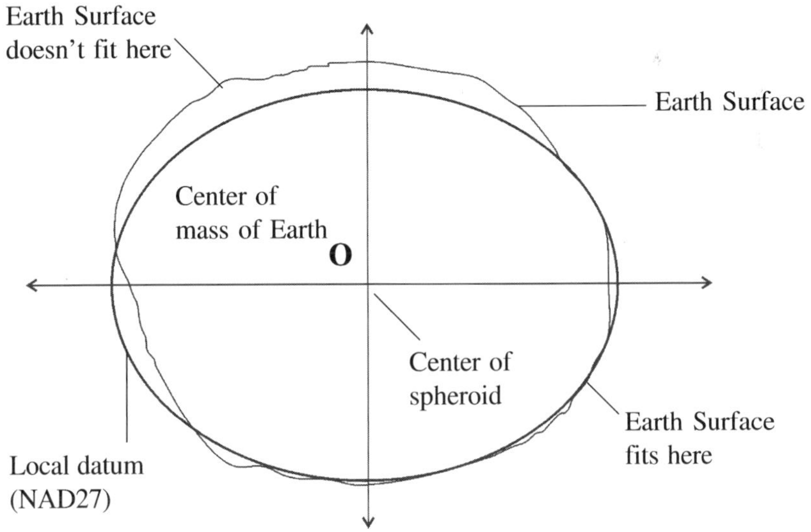

Figure 2.12: Local geodetic datum

2.6.1. Types of geodetic datums

Geodetic datums are usually classified into two categories: local geodetic datums—intended to be used for only a portion of the earth's surface and geocentric datums. A local geodetic datum is best to approximate the size and shape of the particular part the Earth's sea level surface. But this approximation in one region of the Earth is a poor approximation in other parts of the world. Hence the centre of the spheroid will not coincide with the Earth's centre of mass (Fig. 2.12).

Until very recently, most of the country's spatial information systems were based on local geodetic datums. With the introduction of geocentric datum the best approximation of the size and shape of the Earth as a whole could be made. Here the centre of the spheroid coincides with the Earth's centre of mass (Fig. 2.13). Instead of a particular part of the Earth, these datum's application lies in projects or undertakings, which have global applications. The Global Positioning System (GPS), which is operated by the US Department of Defence, utilizes a geocentric datum (World Geodetic System 1984,WGS-84) to express its positions because of its global extent. The Russian GLONASS satellite navigation system also uses a geocentric datum, which is different to that used by GPS. Likewise hundreds of geodetic datums are in use around the world.

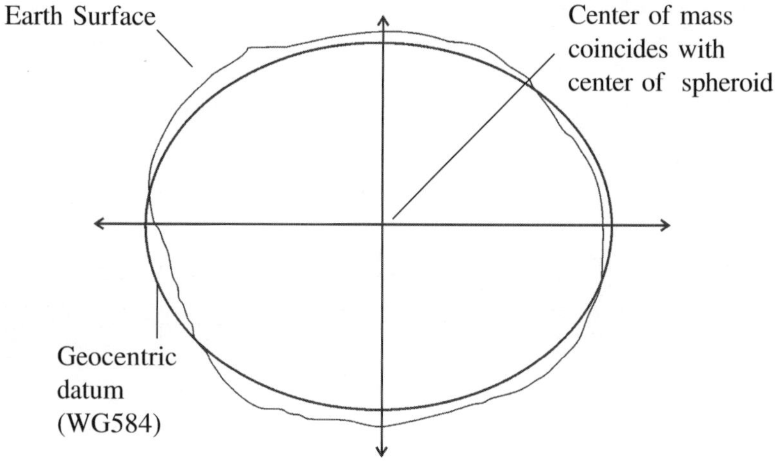

Figure 2.13: Geocentric geodetic datum

Datum conversions are accomplished by various methods. Coordinates can be converted from one datum to another if the relationship between the two is known. The relationship is described by two components, (a) a set of formulae, which describes the mathematics of the transformation process, and (b) a set of parameters, referred to as transformation parameters. The transformation parameters, which are substituted into the formulae, identify the relationship between the particular datums in question. They are calculated by comparing coordinate sets from the two datums. Complete datum conversion is based on seven parameter transformations that include three translation parameters, three rotation parameters and a scale parameter. The Standard Molodensky formulas can be used to convert latitude, longitude, and ellipsoid height in one datum to another datum if the Delta XYZ constants for that conversion are available and Earth Centered Earth Fixed (ECEF) XYZ coordinates are not required.

Multiple maps of the same area must precisely refer to each map. Therefore, it is necessary to convert all maps and datum (in multiple datums) into a single, globally accepted geocentric datum (e.g., WGS-84). Though the mapping organization across the world is striving to achieve this, the process of conversion seems to be very slow and expensive and may take years. Thus, the existence of multiple datums, even for the foreseen future is a reality and one has to be prepared to carefully handle the conversion for the spatial data.

2.7 Map scale and Generalization

Our world is so big that we need to come down to some reduced representation of reality to produce maps. Map is invariably some reductions of the size of Earth features and the 'scale' is an indication of how much it is reduced. 'Scale' can be defined as the ratio of a distance portrayed on the map to the corresponding distance on the ground (Martin, 1996) or an order of magnitude or level of generalization at which phenomena exist or are perceived or observed (Laurini and Thompson, 1992). Some commonly used scales and their corresponding map and ground distance are presented in Table 2.1. Scale can be expressed in one of the three ways, (i) ratio scale; (ii) verbal scale; and (iii) graphical scale (Heywood et. al, 2000).

Table 2.1 Scales of maps

Type of scale	Expression
Ratio	1:250,000
Verbal	1 cm = 2.5 km or 1 cm represents 2.5 km on the ground
Graphical	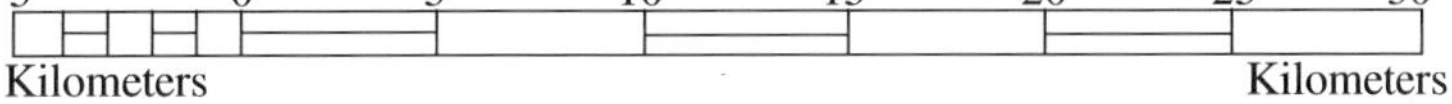 5 0 5 10 15 20 25 30 Kilometers Kilometers

Ratio scale (Table 2.1) is the simple fraction of the distance presented on the map and the corresponding distance as reality. It is also called as representative fraction (RF) method, where both map and ground distances in this have similar units. Thus, at a scale of 1:50,000, a 1 cm line on the map represents 50,000 cm or 500 m or 1/2 km on a ground. Similarly, at the same scale a 1 mm line represents 50,000 mm or 50 m on the ground. Hence, the units are important, but not in the case of *verbal scale*, where one has to clearly mention the units. In *verbal mode,* the scale is expressed as '1 cm = 500 m' or '1 mm = 50 m'. This method of expression is commonly understood by the map user. The third method of expressing scale is the *graphical method.* In graphic mode, a bar (a line) is drawn on the map to illustrate the distance graphically (Table 2.1). Graphical scale is most uncommon and difficult to understand. But it allows an easy transfer from one scale to another. Therefore, graphical scales are mostly used for virtual maps (computer maps) as the user can interactively set the changes quickly.

The scales used in maps vary widely. It may be 1:1,000,000 and 1:10,000. Scales below 1:10,000 are mostly used for cadastral purposes and in managing public facilities and services like electrical, water, sewer, telecommunications, transportation, etc. Map scales can be divided into three categories: small scale maps (scales smaller than 1:1,000,000); medium-scale maps (between

1:75,000 and 1:1,000,000); and large-scale maps (larger than 1:75,000). It is important to remember that small scale maps (e.g., 1:250,000 or 1:1,00,000) cover large areas and large scale maps (e.g., 1:10,000 or 1:25,000) small area. Large scale maps, as compare to small scale maps, contain details of features. A world map with the prominent features like mountains, large rivers, oceans and big lakes are represented in a small scale map. On the other hand, a watershed map with detailed surface features like contours (of small intervals), drains (of different orders), settlements etc., is a large-scale map depicting rich details.

There is no standard scales of mapping as such in a GIS. Depending on the resolution at which data is available and the details at which the objectives of the user are satisfied determines the scale of the GIS map. For example, satellite imageries from LISS III sensor of Indian Remote Sensing Satellite IRS 1D have spatial resolution of 23 m. Any object whose dimension is less than 23 m × 23 m cannot be identified as a separate entity; whatsoever comes within 23 m × 23 m is recorded as one signal (reflectance value). In a study, where remote sensing information is obtained from LISS-III sensor is to be used, other (collateral) information with a spatial resolution of less than 23 m may not be of much help. Similarly, depending upon the grain size of the photographic film, the details of the picture vary. Thus, at every scale, the map shows some degree of generalization. As explained above, even if the data is available at a finer resolution the user may generalize the details in order to bring the clarity of the output image. If the scale is reduced without cutting down the details, it will become too clumsy to interpret. On the other hand, simply enlarging the map (increasing the scale) without putting details of the features cannot reveal much. Using GIS one can always manipulate these for a proper representation of the information at a particular scale. The relationship between scale and details is referred to as *scale-related generalization* (Heywood et. al., 2000).

2.8 Map Projection

If we try to fix a postage stamp on a football it gets fixed without wrinkles. The same if tried for a table tennis ball, the stamp gets wrinkled and it is difficult to fix it properly. In the later case, the area of the stamp is significant in relation to the curvature of the tennis ball, and so it gets wrinkled. Similarly, if we try to cover a ball with a piece of paper, we can't do so without stretching or cutting the paper. This shows that the spherical geometry of the Earth cannot be represented on a flat map without introducing distortion and larger the area, greater is the problem of distortion. However, for every practical purpose we have to project the three-dimensional Earth into a flat surface. Spatial entities (point, line and polygon) must be placed in two-dimensions for effective analyses in GIS, but how?

We know that the latitude and longitude define the position of a point on the surface of the Earth (Fig. 2.7). Similarly a portion or an area of the Earth is defined by a set of latitude and longitude lines. If we are to represent a portion of the Earth surface on a flat paper we are to consider the manner in which we are going to represent the latitude and longitude lines. Systematic arrangement of lines of latitude and longitude on a plane is called map projection. Mathematically, this is the transformation of geographic coordinates (latitude, longitude) into the Cartesian (x, y) coordinates of the map. All map projections can be represented by mathematical equations that convert geographic coordinates into map coordinates.

However, the Earth's graticule (A 'graticule' is a spherical grid of coordinate lines over the planetary surface, comprising ***parallels*** and ***meridians***) has got some important properties like (a) no parallels ever cross one another, (b) parallels are equally spaced along meridians, (c) meridians and parallels intersect at right angles, (d) meridians are equally spaced along parallels, (e) all meridians meet at each geographic pole etc. These properties cannot be very correctly represented on a flat map whatever be the type of projection employed. Projection always introduces distortion of entities. These distortions are conformity, distance, direction, scale, and area (see Box 2.2). Some projections minimize distortions in some of these properties at the expense of maximizing errors in others. Some projections are attempts only to moderately distort all of these properties. But nevertheless one has to carefully examine which properties are correctly represented on the map in order to identify different projections and the inherent patterns of distortion.

Box 2.2

Conformity When the scale of a map at any point on the map is the same in any direction, the projection is caller conformal. Meridians (lines of longitude) and parallels (lines of latitude) intersect at right angles. Shape is preserved locally on conformal maps.

Distance A map is equidistant when it portrays distances from the center of the projection to any other place on the map.

Direction A map preserves direction when azimuths (angles from a point on a line to another point) are portrayed correctly in all directions.

Scale Scale is the relationship between a distance portrayed on a map and the same distance on the Earth.

Area When a map portrays areas over the entire map so that all mapped areas have the same proportional relationship to the areas on the Earth that they represent, the map is an equal-area map.

(*Source*: Dana, 1999)

2.8.1 Classification of map projection systems

Theoretically there may be innumerable map projection systems. However in reality, around 400 map projections are referred in the literature and few dozens of them are used practically. These projection systems possess one or the other distinct properties. These properties are so complex that it is virtually impossible to have a unique rational classification system that group the projection systems uniformly based on these properties. The characteristics of each projection make them useful for some applications and not useful for other applications. However, based on certain criteria, the projection systems can be classified as given in Table 2.2 (Tyner, 1992; Dana, 1999; Dent, 1999; Campbell, 2001).

Table 2.2 Classification of map projection systems

Basis	Classes
1. Method of construction	1. Perspective 2. Non-perspective
2. Preservation of qualities	1. Homolographic/Equal area 2. Orthomorphic/Conformal
3. Developable surface area	1. Cylindrical 2. Conical 3. Azimuthal/Zenithal 4. Conventional
4. Position of tangent surface	1. Polar 2. Equidistant/Normal 3. Oblique
5. Position of viewpoint or light	1. Gnomonic 2. Stereographic 3. Orthographic 4. Others

Before we start discussing different types of map projections, we need to understand some relevant terms.

2.8.1.1 Perspective and Non-perspective Projections

As the names suggest, projections can practically be made by placing a light source inside a transparent globe and project the shadows of the meridians, parallels and other geographic features on a flat surface (like a white sheet of paper) placed tangent to the globe (Fig. 2.14).

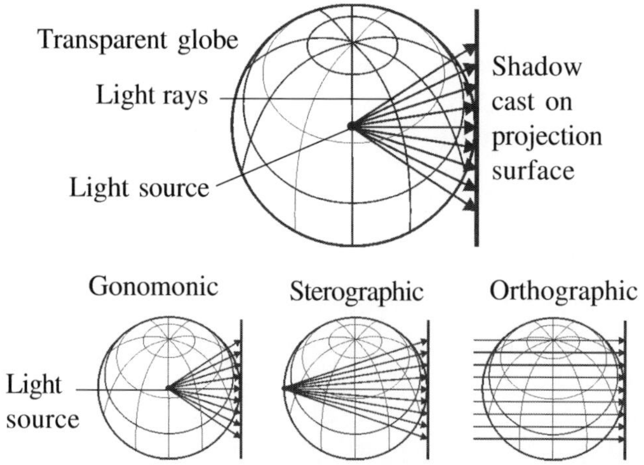

Figure 2.14: Perspective map projection

Here we obtain a shadow image what is called as the 'projection' of the globe. This projection is called *perspective projection*, as against *non-perspective* one where projections are made though mathematical construction. Change the light source and we get different modes of perspective projections. Gnomonic projections place the light source at the centre of the globe, stereographic projections place the light source at the antipode of the point of tangency, and orthographic projections place the light source at an infinite distance from the point of tangency, resulting in parallel light rays. Changing the position of the light source alters the pattern of parallels and meridians on the map, resulting in maps that have different geometric properties.

On the globe the meridians and parallels are circles. When they are transferred to a plane surface, they become intersecting lines, curved or straight. If we stick a flat paper over the globe, it will not coincide with it over a large surface without being creased. The paper will touch the globe only at one point, so that the other sectors will be projected over plane in a distorted form. The projection with the help of light will give a shadowed picture of the globe, which is distorted, in those parts which are farther from the point where the paper touches it. The amount of distortion increases with the increase in distance from the tangential point. But only a few of the projections imply this perspective method. The majority of projections represent an arrangement of lines of latitude and longitude in conformity with some principles so as to minimize the amount of distortion. With the help of mathematical calculations true relation between latitude and longitude is maintained. Hence, various processes of nonperspective projections have been devised. Thus, based on the method of construction, map projections may be classified as *perspective* and *non-perspective* systems.

2.8.1.2 Sequence of Transformations

This is the theoretically step-wise process of producing a flat map of the Earth. The first step is to model the Earth by a simpler solid object (reference globe) having the same surface area as that of the Earth (Fig. 2.15). For large scale maps that show a small portion of the Earth's surface but in great details, the Earth is modelled using an ellipsoid since this gives a better approximation to the true shape of the Earth than a perfect sphere. National mapping agencies in different countries around the world use different ellipsoids, adjusting the lengths of the polar and equatorial radii to get the best fit within their region of interest. For small scale maps that show a large area with little detail, a spherical model is used since it is mathematically simple and at small scales, distortion due to irregularities in the Earth's shape can be considered negligible.

Once a suitable solid model has been selected, the next step is to convert (reduce) the size of the model to the desired scale of the map (generating globe). In this process we reduce the scale by changing the representation fraction (RF) for the reference globe, also called the principal scale (principal scale is the ratio of Earth's radius to that of the generating globe). This scale becomes the nominal scale of the map. However, because of the distortions introduced in representing the spherical globe on a flat map, the scale factor, SF (ratio of actual and the principal scale) will vary in the map from place to place. The final step in the process will be to project the Earth's graticule from the generating globe onto a developable surface (the flat surface onto which projection is made, either a plane, cylinder or cone).

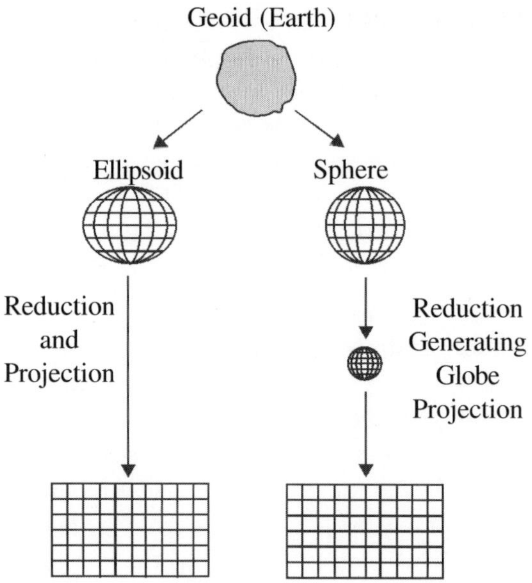

Figure 2.15: Steps in map projection

Thus based on the developable surface, map projections may again be classified as (i) azimuthal, (ii) cylindrical, (iii) conic and (iv) conventional or pseudo-cylindrical. The fourth class does not use a conceptual generating globe.

2.8.1.3 Projection Properties

On the globe, places and areas contain the properties of shape, area, direction and distance. A projection may retain one or two of these but not all, and specifically not both area and shape. A projection that maintains 'area' is called *homolographic or equal area projection*. This means that two equal areas on the globe are also equal on the map. This is achieved by sacrificing shape (stretching in one direction to counter for the Earth curvature must be compensated by compaction in the other). Projections that preserve the shape, such as a square (at the cost of area) is called *conformal* or *orthomorphic projection*. For example, a 2×2 square becomes a 1×1 or 4×4 square. Stretching in one direction is matched by stretching in the other. A projection can thus never be both *equal-area* and *orthomorphic*. Thus, depending upon the preserved quality, a map projection can either be *homolographic* or *orthomorphic*. Projections can again be divided into polar, equatorial or oblique, depending on the relative orientation of the developable surface and generating globe.

A sphere can be projected into any of the three flat (two-dimentional) surfaces (that when unfolded make the map) viz., the plane, the cylinder or the cone (Fig. 2.16), the most rational classification where most of the projection systems can be accommodated would be (i) azimuthal (for plane) [Fig. 2.16 (c)] (ii) cylindrical (for cylinder) [Fig. 2.16 (a)] and (iii) conic (for cone)

[Fig. 2.16 (b)]. These three classes can also be called as projection families (DeMers, 2001), where each specific method becomes a member of the respective families. (For details one can refer to Snyder 1987; Craig, 1882; and Lee, 1944).

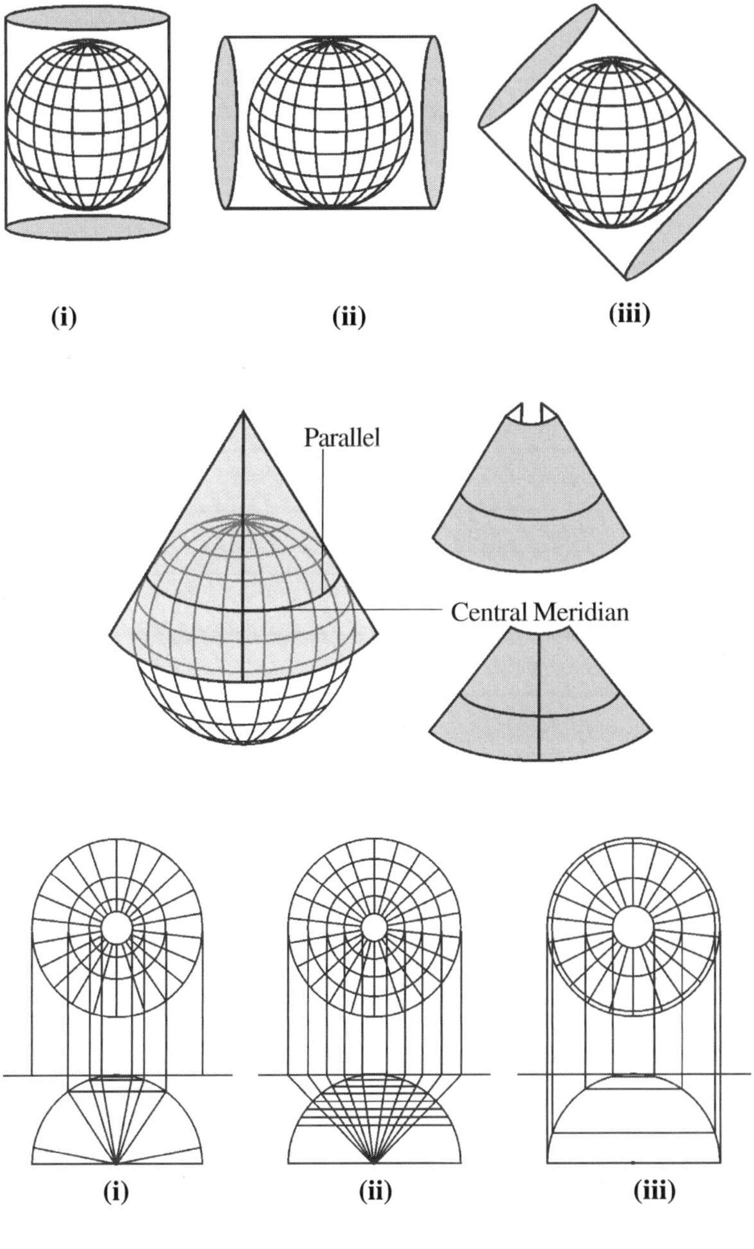

Figure 2.16: (a): Cylindrical projection (i) normal, (ii) transverse and (iii) oblique (b): Conic Projection (c): Azimuthal projection (i) gnomonic, (ii) stereographic (iii) orthographic

2.9 PLANAR OR AZIMUTHAL PROJECTION

This projection can further be classified into perspective and nonperspective projections (Fig. 2.17). The best illustration of a perspective projection is the camera. This form of projection makes the object pass through the optical centre of the lens, forming an image on the screen. There is one to one correspondence between the object point and the image point through the optical centre, which is also the perspective centre. Depending on the placement of the light, the perspective azimuthal projection may be *gnomonic* (the light is inside the globe), *stereographic* (the light is antipodal) or *orthographic* (if light is at infinity) (Fig. 2.14).

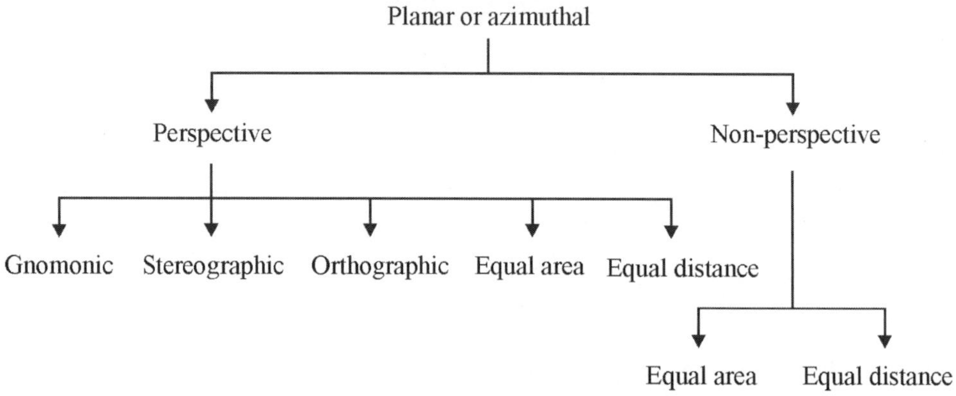

Figure 2.17: Classification of azimuthal projection systems

2.9.1 Gnomonic projection

This projection is obtained by placing a plane tangential to the globe at one point (O) and projecting a point on the globe (P, Q, R, S) by joining it to the centre of the sphere and extending the line to meet the plane (P, Q, R, S), (Fig 2.18).

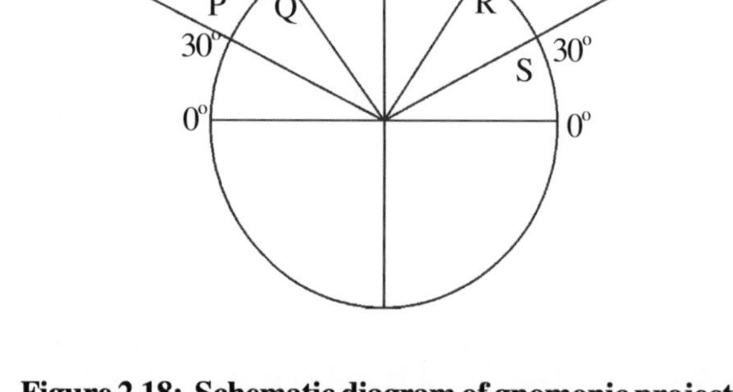

Figure 2.18: Schematic diagram of gnomonic projection

Let a point P is transformed or projected to P′ (Fig. 2.19a) If λ is the latitude of the point P and the radius of the slope is r, then arc NP = OP $\angle$ NOP = $r\,(90° - \lambda)$. Similarly, NP′ = ON tan $(90° - \lambda) = r \cot \lambda$. Hence, in a polar gnomonic projection,

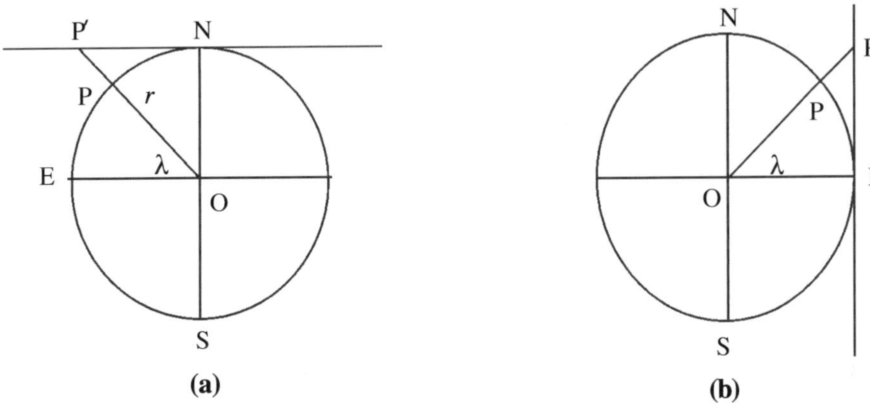

Figure 2.19: (a) Polar gnomonic; (b) Transverse gnomonic projection

- All points on the parallel of the latitude will go as concentric circles of radii r cot λ with N as the centre.
- All meridian will project as straight radial lines, radiating from the pole N at their true angular distance from each other.
- As we go from the north pole towards the Equator the scale will increase and become infinity when $\lambda = 0$.
- The direction to all points will be from the centre or the origin will be absolutely true.
- The complete hemisphere cannot be projected.
- Near the point of contact the distance is reasonably accurate.
- It is easy to develop and used mainly to get the shortest distance between two point from the projection for navigational purpose.

The limitations of this method are that the Equator cannot be projected and large areas cannot be mapped satisfactorily. The use of gnomic projection lies in getting the shortest distance between two points from the projection, which can conveniently be adopted for navigational charting.

2.9.1.1 Transverse or equatorial case

This projection is different than that of polar one in view of that the relationship between P and P′ (Fig. 2.19b) is given by EP′ = OE tan $\lambda = r \tan \lambda$. Thus, the spacing of latitude lines is not linear. As $\lambda \to 90°$, $\tan \lambda \to \infty$, hence the poles cannot be projected. This projection is suitable for regions symmetrical with respect to the Equator. It has advantages similar to those of the polar case but its construction is difficult.

2.9.1.2 Oblique case

This is more complex than the polar or transverse projection. Here the tangent plane is coincident at a point on the globe other than the pole or the Equator. Here again the meridians get projected as straight lines but the parallels go as curved lines.

2.9.2 Stereographic projection

This map projection is obtained by projecting point P on the globe (surface of the Earth) from Earth's south pole (S) to point P′ in a plane tangent to the north pole N (Weisstein, 2004) (Fig. 2.20). Here again we have three cases similar to gnomonic projection as the plane is touching at the pole, Equator or any other point. Geometrically, if one sits on a plane and thinks Earth as a transparent globe, one (the south pole is a point on the plane) can visualize a light source at the north pole that comes through the sphere. Each point on the sphere then casts a shadow on the plane which can be drawn to get the projection of the Earth (Sullivan, 2004). Hence the south pole appears to be the center point on the map, around which the lines of latitudes appear as circles. The northern hemisphere will be stretched as compared to the southem hemisphere. Even then let P be a point on the globe of latitude 30° and P′ be the position on the plane which is projected by joining with S.

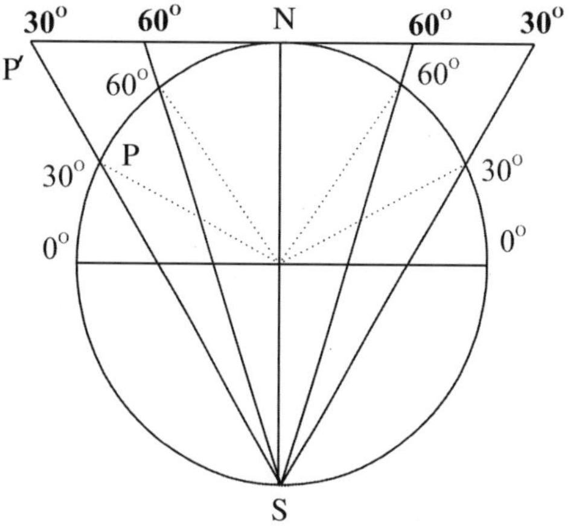

Figure 2.20: Stereographic azimuthal projection

In this system, the scale exaggeration along the parallels is not as much as for the gnomonic projection. The meridians are projected at correct spacings. As the circles map to circles in the plane, the angles are preserved. Thus the stereographic is conformal. As this projection has a simple algebraic form, it is also easy to construct. The meridians and parallels projected on the plane also interest at right angles.

2.9.3 Orthographic projection

In this projection the perspective centre recedes to infinite and hence all the projecting rays are parallels (Fig. 2.21a). Here again we can have the tangential plane at the pole, equator, or any other point giving rise to polar, equatorial and oblique cases.

Referring to the Figure 2.21 (b), a point P is projected to P′, so that $NP' = PC = r \cos \lambda = r \sin (90° - \lambda)$ and $NP = r (90° - \lambda)$. For small values of $(90° - \lambda)$, this is the point near the pole, $NP' \cong R (90° - \lambda)$. As we go away from the pole there is a general reduction of the scale on the orthographic projection because NP′ will be less than the arc distance NP. However, the scale along the parallel is always true because $P'N = PC$ (for any latitude). This is a projection from infinity and it preserves neither area nor angle (Weisstein, 2004). But an advantage of this system is that it is possible to apply a uniform coordinate system like grid to the projected image and also to 'scale' them maintaining the exact proportions and measurements. This projection can mainly be used for illustration, as it shows the Earth as seen from space infinitely far away. Severe shape and area distortion near the map bordes present general use of this projection for creating world maps (Furuti,1997).

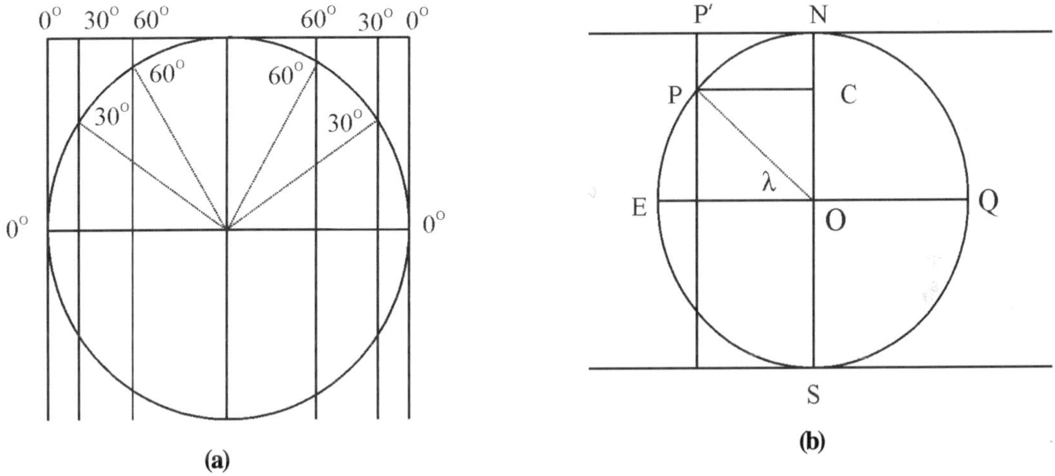

Figure 2.21: (a, b) Orthographic azimuthal projection

Perspective equidistant projection The property of equidistant in a perspective projection can be obtained by moving the perspective centre to a certain specified point away from the globe and getting the projection on a tangential plane.

Perspective equal area projection In this case, we shift the perspective centre to a point such that projected points will give equal area as the globe. In this type of projection, the meridians and the parallels are projected as curved lines.

2.9.4 Nonperspective projection

In the perspective projections discussed so far, the point of view or the perspective center remain fixed. However in the nonperspective protection projective center is moved from place to place to achieve certain properties, i.e., equal area or equal distance. These unlike 'classical' or perspective projections are purely mathematical in nature. A full sphere is mapped with an inner hemisphere surrounded by a ringlike outer one (Furuti, 1997).

2.9.4.1 Equidistant polar zenithal projection

Here the parallels are drawn as concentric circles of radii $r\,(90° - \lambda)$ because NP = $r\,(90° - \lambda)$,

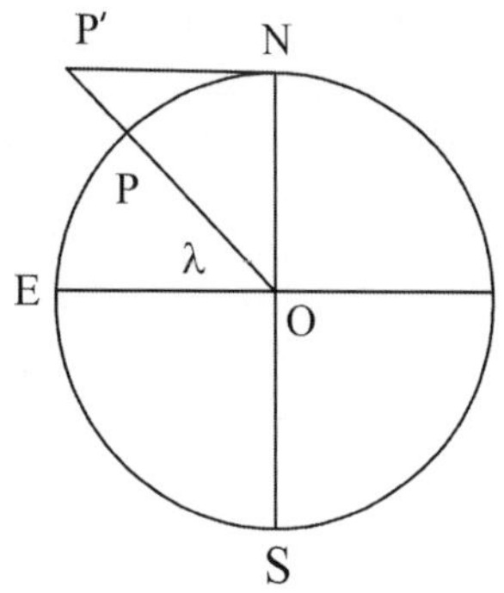

Figure 2.22: Equidistant polar projection

so that for any point of latitude NP = NP′ = r (90° – λ). The meridians are (Fig. 2.22) also drawn equally spaced as straight radial lines. Distances are thus along the meridian but not true along the parallels of latitude. This can be concluded from the simple fact that length of the parallels on globe is $2\pi r \cos \lambda$ whereas on this projection it is $2\pi r$ (90° – λ). Since $2\pi r \cos \lambda$ is always less than $2\pi r$ (90° – λ) there is an exaggeration of scale along the parallels of latitude on this projection and consequently distortion of shape and area from the centre of projection is observed. Near the centre of the projection, distortion is small for shape, area and distance and hence the projection is satisfactory. The construction of the projection is easy, as we have to draw only concentric circles and radial lines.

2.9.4.2 Equal area zenithal projection

In this system, the quadrilaterals formed by meridians and parallels have the area on the map proportional to their area on the globe. The projection compensates the increase in scale along parallels with an appropriate decrease in scale along meridian.

2.10 Cylindrical Projection

Under this group of projection, the surface on which projections are obtained is a cylinder (Fig. 2.16(a). As explained in Figure 2.23, a cylindrical projection of points on a sphere (with unit radius) consists of extending the line OS (O being the centre of the globe) until it intersects a cylinder tangent to the sphere at its equator at a corresponding point A. If the sphere is tangent to the cylinder at longitude λ_0, a point on the sphere with latitude ϕ and longitude λ will be mapped to a point on the cylinder with height tan ϕ. Unwrapping and flattening out the cylinder gives the cartesian coordinates as

$$x = \lambda - \lambda_0 \text{ and}$$

$$y = \tan \phi$$

The cylinder can be tangential to the globe at the equator or poles, or be in an oblique position. The cylinder can also cut the globe to get two parallels along which the scale is true. Again, we can have perspective and non-perspective or conventional projections. By suitable combinations we can get projection for equal area, equidistant or orthomorphic properties. These projections are most suitable when the map coordinates are to be described in rectangular or grid systems whereas the azimuthal projections are good for polar coordinates. The difficulty of projecting poles in the normal case and that of the equator in the transverse case limits the use of perspective

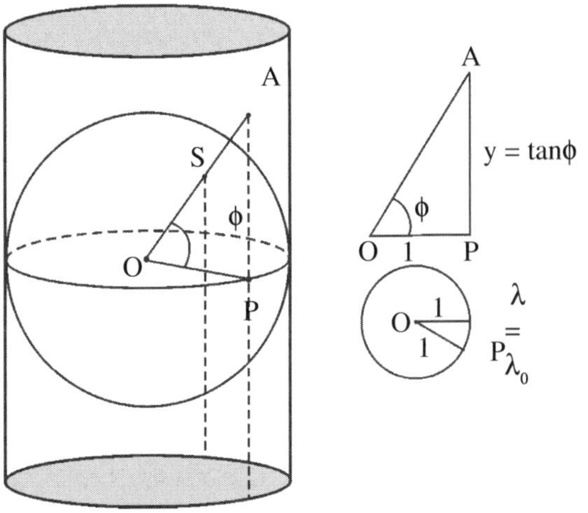

Figure 2.23: Cylindrical projection (source: Weisstein, 2004)

projection. There are a number of non-perspective projections which are well known, the Mercators projection being the most common among them.

Cylindrical projection systems are as such seldom used in practice. However, the term 'cylindrical projection' is instead used to refer to any projection in which lines of longitude are mapped to parallel lines and latitudes are mapped to parallel lines with arbitrary mathematically spaced separations (Snyder, 1987). Nevertheless, some of the important members of this family are briefly discussed here.

2.10.1 Perspective cylindrical projection

This case can be illustrated by taking a normal position of the cylinder when it is touching all along the Equator. Refer to Figure 2.24, we note that a point P will be projected to P´ and the poles N and S cannot be projected at all. The 'normal' case assumes that the cylinder is tangent to the globe along the equator (Fig. 2.16). The meridian will go as straight parallel lines, equally spaced and the latitude lines also as straight lines parallel to and with the same length as the equator. The equator will go as of correct length = $2\pi r$. The scale exaggeration along the λ, parallel of latitude is sec λ. Near the equator $r \tan \lambda \cong \lambda$ and hence the scale is true on the meridian close to the equator. But as we go away from the equator the scale exaggeration becomes much pronounced. Like normal azimuthal projections, normal

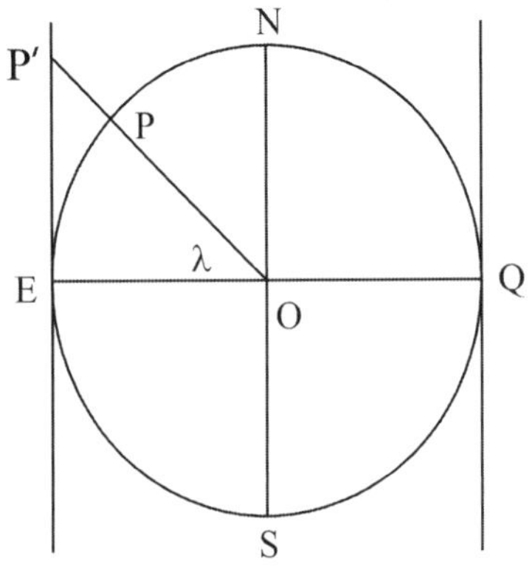

Figure 2.24: Perspective cylindrical projection

cylindrical projections have different geometric proportion stained by modifying the spacing of the parallels.

2.10.2 Cassini projection

This is another transverse case. Here the map coordinates are defined by

$$x = \sin^{-1} [\cos \phi \sin (\lambda-\lambda_0)]$$

$$y = \tan^{-1}\left[\frac{\tan\phi}{\cos(\lambda-\lambda_0)}\right]$$

Actually, it is a transverse aspect of Plate Carree projection—the most simple and limiting form of equidistant cylindrical projection. This projection is suitable for countries, which have a narrow belt in longitude and long extension in latitude. This projection is also suitable for superimposing rectangular grid with easting along the parallels and northing along the meridians. In India this projection is largely used for large scale mapping with the central meridian and 1° extension both in longitude and latitude.

2.10.3 Gall's projection

It is a nonperspective projection in which we have a hollow cylinder intersecting the globe at latitude 45° N and S, the viewpoint taken stereographically and shifting from meridian to meridian (Fig. 2.25). In the immediate vicinity of 45° the scale will be correct. Near the poles, the exaggeration along the parallels of latitude is much greater than the exaggeration along the meridians. As such representation along the poles is unsatisfactory. Even then, it is an improvement over Mercator in removing distortion of sizes at latitudes approaching the poles. As the map stretches vertically, so do the countries on the map. But the standard parallels remain at 45° north and south, the temperate countries are the ones better shown than the tropicals.

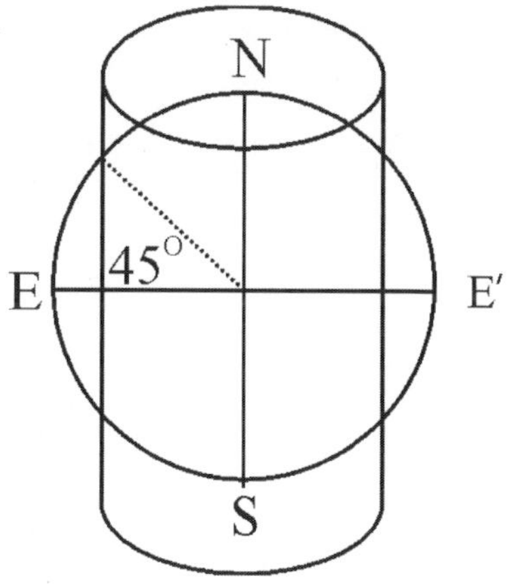

Figure 2.25: Gall's projection

2.11 Conic Projection

Unlike the cylindrical projection, the surface of projection is a cone (Fig. 2.26). For each point S on the globe, the extension of line OS intersects the cone at point C (the globe touches the cone at point T). If h (AO) be the height, the angle from the T = axis at which the cone is tangent is given by $\theta = \sec^{-1}h$. Here again the cone may either touch the globe along a parallel of latitude or cut the globe along two parallels. The parallel of latitude along which the cone touches or cuts is called standard parallel. The scheme of various possible projections is given in the Figure 2.27.

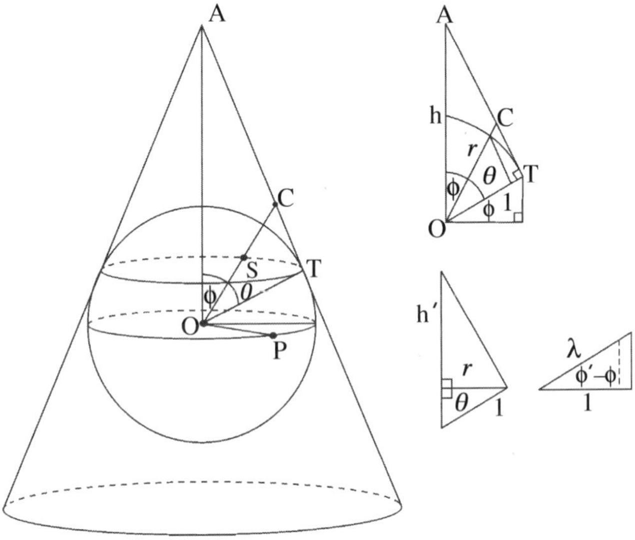

Figure 2.26: Conic projection

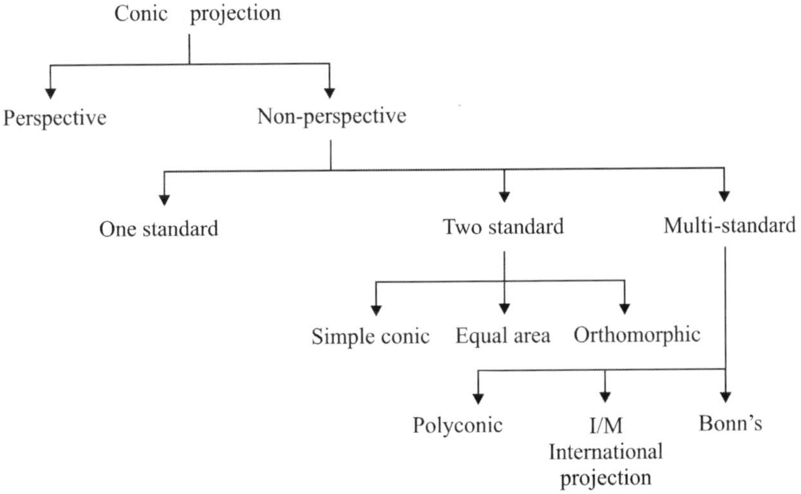

Figure 2.27: Classification of conic projection

2.11.1 Simple perspective conic projection

Let us consider a cone whose axis coincides with the axis of the globe and let it touch a parallel of latitude. The cone when developed will be as shown in (Figure 2.28). Let C be the apex of the conic. To find the angle of cone at C, we notice that

$PP' = 2\pi r \cos \lambda$ (circle of the radius OP)

$CP = r \cot \lambda$ (from right angled triangle OPC)

Therefore, $\angle PCP' = \text{arc/radius} = (2\pi r \cos \lambda) / r \cot \lambda$

$= 2\pi \sin \lambda = 360° \times \sin \lambda.$

The quantity $\sin \lambda$ is called the constant of the developed cone. This projection has all meridians equally spaced. All parallels of latitude are projected as concentric.

2.11.2 Simple conical projection with two standard parallels

In this case the globe is cut in two parallels of latitude l and l'. These two lengths are projected correctly and the scale is true neither along these parallels, but neither inside nor away from the two standards. To construct the projection we have to calculate the radii for the two parallels on the developed cone and also its cone angle or the apex angle (Fig. 2.29). This projection has the advantage that the scale is true along the meridians and along the two standard parallels. This increases the area of greater accuracy. This projection can be visualized by introducing a negative scale at the central meridian. The scale decreases between the two standard parallels but increases outside the two standard parallels. The projection is suitable for narrow belts of latitude but the range of two standard conical projection is much more than of one standard projection.

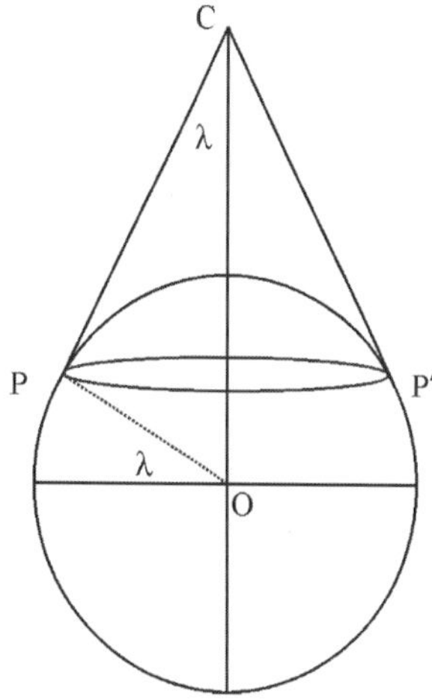

Figure 2.28: Simple perspective conic projection

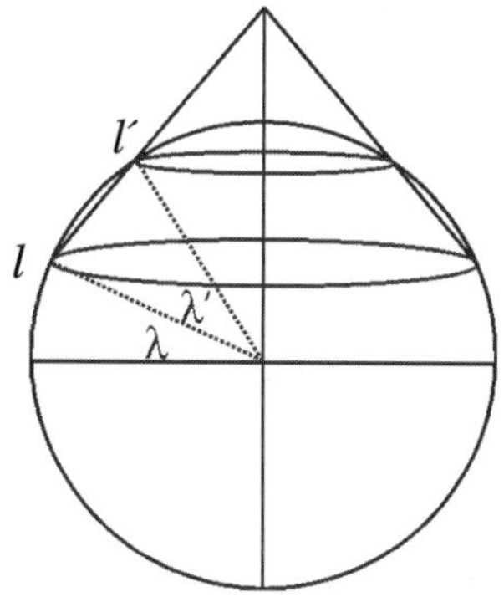

Figure 2.29: Simple conic projection with two standard parallel.

2.11.3 Equal area conic projection with two standard parallels

In this conic projection it distorts scale and distance except along standard parallels. Areas are proportional and directions are true in limited areas. It is used in the United States and other large countries with a larger east-west than north-south extent.

2.11.4 Orthomorphic conic projection with two standard parallels

In this projection area and shape of the projected map are distorted away from standard parallels. Directions are true in limited areas. It is in use for maps of North America.

2.11.5 Equal distant conic projection with two standard parallels

In this particular projection, direction, area, and shape are distorted away from standard parallels but the scale remain constant. It is used for portrayals of areas near to, but on one side of the equator.

2.11.6 Polyconic projection

It is another member of conic projection system (Fig. 2.36) where parallels are represented by a system of non-concentric circular arcs with centers lying on the straight line representing the central meridian (Lee, 1944). The coordinats are defined by

$$x = \cot\phi \sin\{(\lambda - \lambda_0)\sin\phi\}$$

and $$y = (\phi - \phi_0) + \cot\phi\,[1 - \cos\{(\lambda - \lambda_0)\sin\phi\}]$$

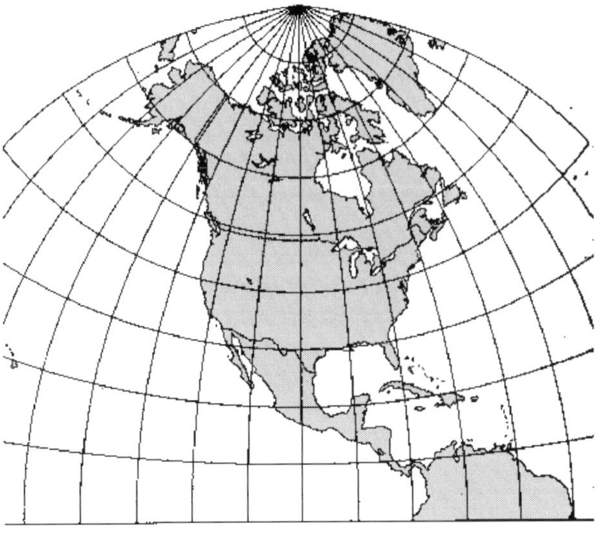

Figure 2.30: Polyconic projection

This projection can be used for topographical maps for small areas but not for large ones. In fact for topographical mapping the polyconic projection is slightly modified by shifting the central meridian from place to place as the scale is independent of the central meridian when a number of such topographical sheets are joined together (Fig. 2.30) they fit nicely in north-south direction but in east-west direction they have a kind of rolling fit.

2.11.7 International maps (IM projection)

This is a modified polyconic projection. The individual covers 4° of latitude and 6° of longitude. The first modification concerns in making the scale along each meridian of 2° from the center true. As a result of the scale along the parallels become slightly smaller because parallels are still drawn with radius $R \cot \lambda$. The second adjustment is also about the meridians. The meridians are drawn as straight at their intersection points with the parallels. This allows the joining of the individual sheets in the east west direction, thus minimizing the effect of rolling fit of polyconic projection.

2.11.8 Bonn's projection

It is a pseudo-conic projection and is constructed by starting up with a simple conic projection in which the parallels are equally spaced along the meridians and then scaled to be true to scale. A central meridian represented by a straight line is selected. A central parallel (center of the area) is selected as a standard parallel. The standard parallel is drawn with its proper radius as if developed from its own cone. This limits binding the meridians such that they become a true scale. The central meridian is divided true to scale for the intersection of parallels. The other parallels (Fig. 2.31) are drawn as concentric circles with N as center and true to their lengths by subtending true angles at the center of the projection. This projection has the following properties:

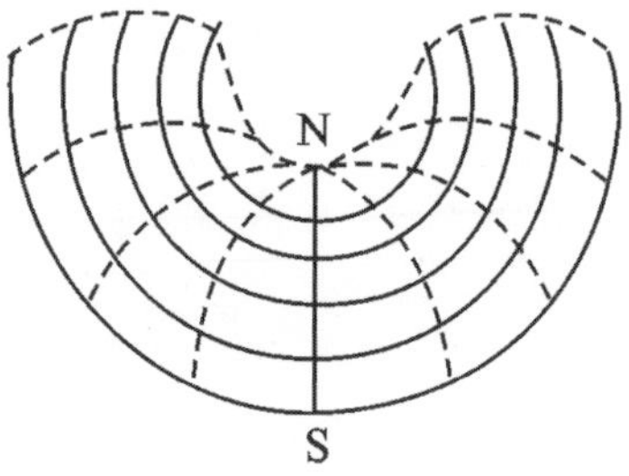

Figure 2.31: Bonn's projection

- The scale is correct along the central meridian and other parallels.
- The intersection of meridians with all other parallels and of all other meridians with the standard parallels are rectangular.
- Scale gets exaggerated away from the central meridian.
- Projection is an equal area because the parallels are of true length. Here a rectangle goes over as a parallelogram of equal area.
- Representation is very good near the central meridian.
- Projection is suitable for large latitude belt but small longitudinal extent like South America.

2.12 SELECTION OF A PARTICULAR PROJECTION

In view of large number of available projections, the cartographer is sometimes faced with a difficult task of selecting a suitable map projection for his work. His decision is based on the following considerations:

(i) The function and purpose of the map;

(ii) Area, general shape and position or location on the surface of the earth of the region to be mapped;

(iii) Extent of the area to be represented i.e., whether the sheets are to be independent or the area is to be contained in one sheet. This however, may include a number of sheets, which form part of one whole sheet. For example, in polyconic projection, every sheet is projected with respect to its own meridian independently and hence, the individual sheets cannot be fitted together.

Another minor consideration sometimes is the case of construction of the projection. In any projection, there are three fundamental properties. The important categories, which affect the first consideration, are those of the economist, politicians and geographers. The first one is interested with distribution of patterns, and hence to him the area of relationship comes first. As a second consideration he may be interested in the correct distance and direction relationship also. The second category involves the guiding, directing or recording of motion. In this category will come the mariner, the meteorologists, the navigators, the civil engineers and army personnels. They are interested in measurement of angles and distance. For this they require orthomorphic projection and projection of equidistant groups. The third category of people, namely, the geographers, is interested in the geographic records of the Earth surface. They are not concerned with minor distortion and slight difference of scale. However the map projections are usually applied for the following purposes:

(a) Computation of world maps and atlas maps.

(b) Computation of grid coordinates in map as a base for topographical maps.

(c) Computation of map coordinates when geographical coordinates are available.

CHAPTER 3

Maps as Numbers: Data Models and Data Structure

"Yes raster is faster, but raster is vaster, and vector just seems more corrector."

(C. Dama Tomlin, *Geographic Information Systems and Cartographic Modeling*, 1990, p. 44)

3.1 MAPS AS NUMBERS

To a common man a map is simply some drawings on a paper to guide him through some unknown new places. These traditional maps are available in many forms and sizes, but in all the cases, the medium is always the paper. On the contrary geographical information systems require that a map (with all its contents-text, drawings, symbols, digits) has to be represented by numbers. The organization of maps into digits is significantly important in the context of our capturing, handling and analyzing map data in a GIS. This Chapter deals with how the paper maps (or in other words, the real world) can be viewed as digits (data) and how we can link the features we are observing in real sense and the actual files (bits and bytes) inside the computer (GIS).

Both the data and graphics (maps) have to be stored in a GIS as numbers (digits). The data is stored in the computer memory as a physical data structure (i.e., files and directories). Here we are more interested in not the way in which computer memory is used (like disk or RAM) but the mechanism by which the files (and directories) store and access the spatial (map data) and the attribute information. The files may be written in *binary* as well as *ASCII* (Pronounced 'ask-ee') code. The former is smaller and faster to read and the latter takes more space. The map, just like attributes, can be dispersed into sequential number pieces to be stored in the computer memory. When the numbers are stored in files in binary mode, we refer to 'bits' (binary digits). Eight bits in a row is called a 'byte'. One byte can represent numbers from 0 (0000 0000) to 255 (1111 1111). Several bytes are linked together to make a 'computer word', which is grouped into 'data record'. Several data records group together to form a 'computer file' and then sets of files form sub-directories and directories. Detailed information about bits, bytes and the binary system are presented in Box 3.1 and 3.2.

Box 3.1 Bits and Bytes

Bit It stands for 'binary digit'—the smallest unit of information in a digital world, introduced by John Tukey, an American statistician and computer scientist in 1946. A bit is represented by the numbers 1 and 0, which correspond to the states on and off, true and false, or yes and no. Bits are the building blocks for all information processing that go in digital electronics and computers.

Byte It is built from bits. One byte equals 8 bits. This term was first used in 1956 by an American computer scientist Werner Buchholz to avoid confusion with the word bit and was soon adopted by the computer industry as a standard. The values that a byte can take range between 00000000 (zero) and 11111111 (255). This means that a byte can represent 28 (28) or 256 possible states (0-255).

Bytes are combined into groups of 1 to 8 bytes and are called words. The size of the words used by a computer's central processing unit (CPU) depends on the bit-processing ability of the CPU. A 32-bit processor, for example, can use words that are up to four bytes long ($4 \times 8 = 32$ bits). The number of bits used by a computer's CPU for addressing information represents the computer's speed and power. Computers today often use 16, 32, or 64 bits in groups of 2, 4, and 8 bytes. Computer graphics are described by the number of bits used to represent pixels—the smallest identifiable parts of an image. In monochrome images, each pixel is made up of one bit. In 256-color and gray-scale images, each pixel is made up of one byte (eight bits). In true color images, each pixel is made up of at least 24 bits. The particular sequence of bits in a byte encodes a unit of information such as a keyboard character. One byte typically represents a single character such as a number, a letter, or a symbol.

Bytes are the major unit for measuring quantities of data as kilobytes (1024 103 bytes), megabytes (1,048,576 106 bytes), or gigabytes (about 1 billion 109 bytes). A regular floppy disk normally holds 1.44 megabytes of data, which equates to approximately 1,400,000 keyboard characters. At this storage capacity, a single disk can hold a document approximately 700 pages long, with 2000 characters per page.

The other way to code the map data into computer files is called ASCII (American Standard Code for Information Interchange). The ASCII is also called 'text code' as this format is similar to what we deal with normal text, punctuation etc., ASCII codes may be numbers, letters (both lower and upper cases), some special characters like $ punctuations and also keys to operations (tabs). In an ASCII file, each alphabetic, numeric or special character is represented with a 7-bit binary number (a string of seven 0s or 1s). There are 128 standard ASCII codes each of which can be represented by a 7-digit binary number: 0000000 through 1111111, plus parity. A detailed account of ASCII code is presented in Box 3.3.

Box 3.2 Binary System

In computing, **digital** is synonymous with binary. One bit can represent at the most two values; 2 bits, four values; 8 bits, 256 values; and so on. Values that fall between two numbers are represented as either the lower or the higher of the two.

The main system of mathematical notation is the decimal system (a base-10 system). While in decimal system the base is 10, in binary system it is only two digits (on and off or 1 and 0) to represent a number. Any decimal number can be converted into the binary number by summing the appropriate multiples of the different powers of two. In binary system a number of four bits like 0101 represents 5 (five) in decimal system in the following way:

0101 (in binary) = (23 * 0) + (22 * 1) + (21 * 0) + (20 * 1) = 510 (in decimal). Similarly, 10101101 represents (1 × 20) + (0 × 21) + (1 × 22) + (1 × 23) + (0 × 24) + (1 × 25) + (0 × 26) + (1 × 27) = 173. Likewise, the first 20 numbers in the binary notation will be:

1: 00001; 2: 00010; 3: 00011; 4: 00100; 5: 00101; 6: 00110; 7: 00111; 8: 01000; 9: 01001; 10: 01010; 11: 01011; 12: 01100; 13: 01101; 14: 01110; 15: 01111; 16: 10000; 17: 10001; 18: 10010; 19: 10011; 20: 10100

For the conversion of decimal numbers to binary numbers, the same principle can be used, but the other way around. Here, we first find the highest power of two that does not exceed the given number, and place a '1' in the corresponding position in the binary number. For example, the highest power of two in the decimal number 519 is 29 = 512. Thus, a '1' can be inserted as the 10th digit, counted from the right: 1000000000. In the remainder, 519 – 512 = 7, the highest power of 2 is 22 = 4, so the third zero from the right can be replaced by a

1: 1000000100. The next remainder, 3, consists of the sum of two powers of

2: 21 + 20, so the first and second zeros from the right are replaced by

1: 519 = 10000001112.

Arithmetic operations in the binary system are almost similar to decimal system with the basic rules as: 1 + 1 = 10, and 1 × 1 = 1. Zero plays its usual role: 1 × 0 = 0, and 1 + 0 = 1. Addition, subtraction, and multiplication are done in a fashion similar to that of the decimal system:

100101	1011010	101
+110101	– 110101	×1001
1011010	100101	101
		0000
		00000
		101000
		101101

Because only two digits, or states (on and off, 0 and 1) are involved, the binary system is useful in computers, which are digital devices. The "on" position corresponds to a 1, and the "off" position to a 0. In magnetic storage devices (hard disk, floppy, zip, tape etc.,) magnetized areas of the media are used to represent binary numbers: a magnetized area stands for 1, and the absence of magnetization means 0. Flip-flops—electronic devices that can carry only two distinct voltages at their outputs and can be switched from one state to the other state by an impulse—can also be used to represent binary numbers; the two voltages correspond to the two digits. Optical and magneto-optical storage devices use two distinct levels of light reflectance or polarization to represent 0 or 1.

Source: http://www.usbyte.com/common/Binary.htm

Using GIS, we need to generate an accurate, registered, digital map that can be queried and analyzed. The real world paper map has to be translated into computer files, using some spatial data models (a data model is the conceptual description of how data are organized for use by the GIS) with the entity information being translated into queriable database files using relational or object-oriented database. Thus, the spatial coordinates are related with entity information.

Box 3.3 ASCII System

ASCII stands for American Standard Code for Information Interchange. As computers can only understand numbers, so pan ASCII code is a numerical representation of an English character with each letter assigned a number from 0 to 127. For example, the ASCII code for uppercase 'A' is 65 and for lowercase 'a' is 97. Likewise ASCII code for '+' sign is 43 and for the '–' is 45. As most of the computers use ASCII codes to represent text, it is possible to transfer data from one computer to another. The text means 'plain text' with no formatting like bold, underscore, tabs etc., As this raw form is supported by every application in every machine and thus can be imported and exported to any application. Notepad.exe creates ASCII text; similarly any file in MS Word can be saved as 'text only'.

Text files stored in ASCII format are sometimes called ASCII files. Text editors and word processors are usually capable of storing data in ASCII format, although ASCII format is not always the default storage format. Most data files, particularly if they contain numeric data, are not stored in ASCII format. Executable programs are never stored in ASCII format. The standard ASCII character set uses just 7 bits for each character. There are several larger character sets they use 8 bits, which give them 128 additional characters. The extra characters are used to represent non-English characters, graphics symbols, and mathematical symbols. Several companies and organizations have proposed extensions for these 128 characters. The DOS operating system uses a superset of ASCII called *extended ASCII* or *high ASCII.* A more universal standard is the ISO Latin 1 set of characters, which is used by many operating systems, as well as Web browsers. One can visit *www.asciitable.com* for a complete list of commonly used characters and their ASCII equivalents.

3.2 Abstract Representations of Reality

Lets have a look around. What do we see? When asked to describe what is being seen, we speak of many objects that fall in our view. These may be trees, buildings, road, fields, or landscape in general or even tables, chairs, bookshelves and so on. We can call them as 'units' or 'objects'. But there is something more to know, i.e., 'where it (the unit/ object) is'? It refers to the exact or relative position (because a common man never describes the positions in terms of latitude/ longitude or XY coordinates; rather the position with relation to other things as 'near by' or 'at the side of', 'in front of', 'beside' or 'to the right of' etc. of the objects. In GIS, this is called as 'geo-referencing'. Nevertheless, one person's interpretation of the real world, which he/she is observing is likely to differ from other person depending on individual's experience, socio-economic and cultural background and so on. These perceptions of the real world and their descriptions are the 'conceptual' models and there may be as many as conceptual models as the number of the people in the world.

Thus it becomes imperative to store in a system like GIS the real world phenomena based on some formalized models representing the world. The creation of an analogue or digital representation of human observation of the world involves seven levels of model development and abstraction as given by Burrough (1986) (Fig. 3.1). These are:

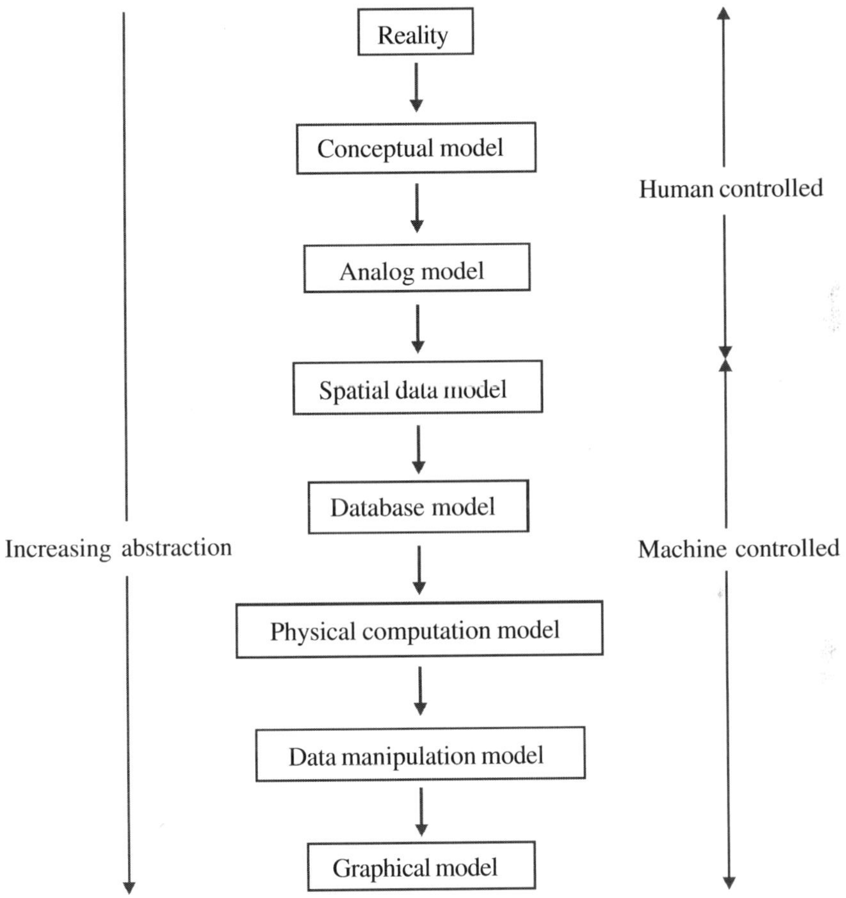

Figure 3.1: Levels of abstraction in relation to model development in GIS

(i) A view of reality (conceptual model)

(ii) Human conceptualization leading to an analogue abstraction (analog model)

(iii) A formalization of the analog abstraction without any conventions or restrictions on implementations (spatial data model)

(iv) A representation of the data model that reflects how the data are recorded in the computer (database model)

(v) A file structure, which is the particular representation of the data structure in the computer memory (physical computation model)

(vi) Accepted axioms and rules for handling the data (data manipulation model)

(vii) Accepted rules and procedures for displaying and presenting spatial data to people (graphical model)

Fundamentally, we adopt two approaches to formalize the conceptual ideas of space and spatial properties. They are – (a) to perceive the space as 'entities' i.e., distinct units of a real world phenomenon (e.g., buildings, roads, wells, etc.) – 'object-based model', or (b) to perceive space as continuous 'fields' (e.g., soil pH, atmospheric humidity, elevation, etc.,) – 'field based model'. In the former case, the space consists of discrete objects described by their attributes (i.e., 'what is') and the exact or relative positions on a map (i.e., 'where is'). To imagine the space as fields involves describing it as continuous features where the attributes vary continuously over space.

3.3 Spatial Data Model

Spatial variation in the real world is infinitely complex. The closer one looks, the more details he/ she can see, almost without limit. Thus, it would take infinitely large data sets to capture the real world at any level of precision. Different types of people have different perceptions. They also use GIS for different purposes and the type of phenomena people study have different characteristics. Hence, it cannot be a single type of data model that is best for all circumstances. On the other hand, computer capability is finite. The data (and the perception) must somehow be uniformly directed and reduced to a finite and manageable quantity by a process of generalization or abstraction. This necessitates representing the (geographical/ spatial) variation in terms of discrete elements or objects. The 'rules' that are used to convert real geographical variation into discrete objects are contained in a *data model*. It is a set of constructs for describing and representing parts of the real world in a computer system (Maguire et al. 1991) or a set of guidelines for the representation of the logical organization of the data in a database… (consisting) of named logical units of data and the relationships between them (Tsichritzis and Lochovsky, 1977).

The *spatial* data model represents a set of guidelines to convert the real world to the digitally and logically represented spatial objects consisting of the attributes and the geometry. The attributes are managed by thematic or semantic structure while the geometry is represented by geometric-topological structure. We assume that any geographical phenomena can be effectively discretized into small parts for analysis, each part can be identified, their attributes measured and the geographical coordinates registered. Thus, the simplest data model of reality is a basic *spatial entity* (basic carrier of information), which is specified by its attributes and the geographical location. A spatial entity can be, according to one of the three basic geographical data primitives, point, line or area. These three basic entity types can be used to formalize any spatial phenomenon or process. We can also refer to two additional spatial entities, namely surface and network. A surface entity is used to represent continuous features or phenomena (unlike to the point, line or areas, which are discrete). A ***network entity*** is a series of interconnecting lines along which there is a flow of data or objects.

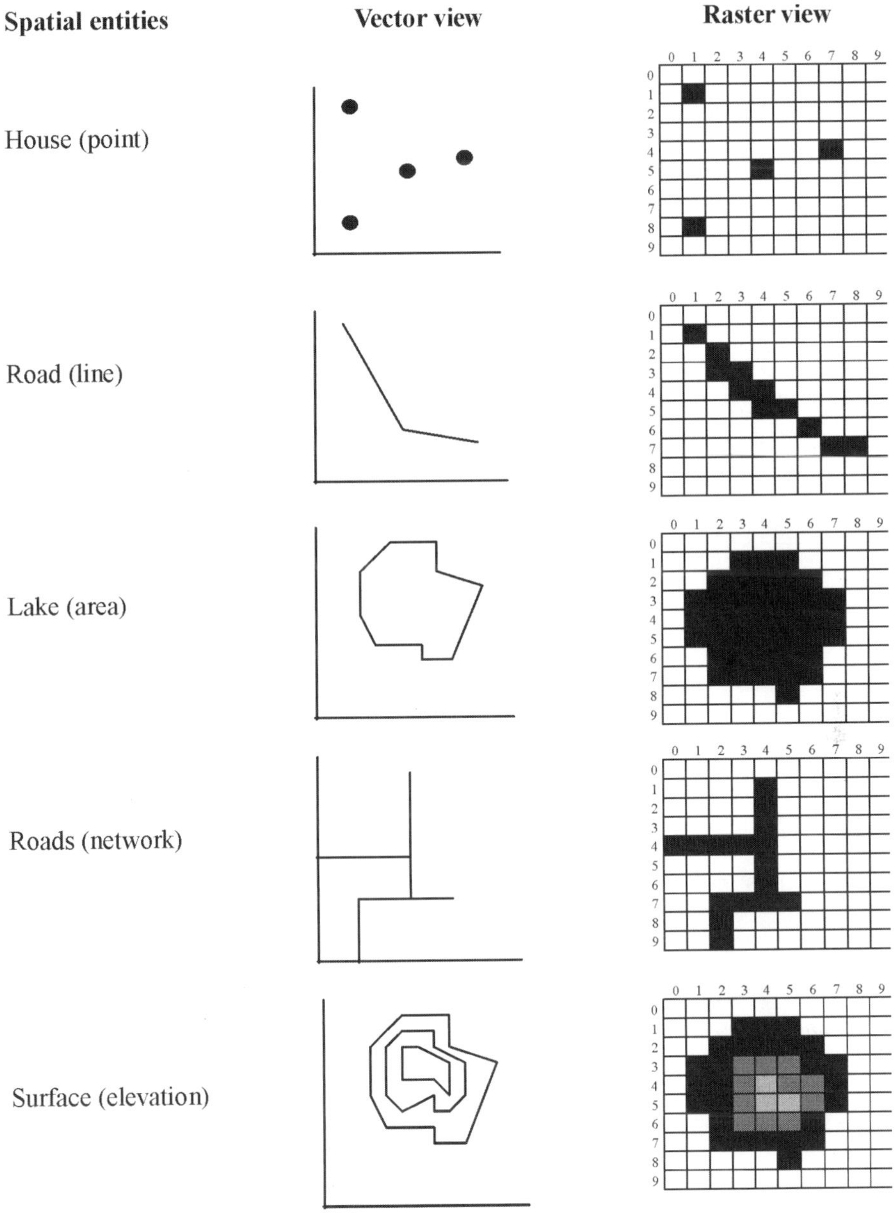

Figure 3.2: Raster and vector spatial data models

Computers require unambiguous instructions on how to handle and display a data model. Two universally accepted models of spatial data model types, vector and raster are used in GIS. Vector model uses discrete points, lines and/or areas corresponding to discrete objects with name or code number of attributes that are georeferenced by two-dimensional Cartesian coordinates (x, y). Raster model is described by tessellations. It uses regularly spaced grid cells, the shape and characteristics of which determines the entity types. An element of the grid cell is called a 'pixel' (picture element or picture cell). The conventional sequence is row-by-row from left to right and then line-by-line from top to bottom. Every location is given in two dimensional image coordinates; pixel number and line number, which contain a single value of attributes. Figure 3.2 depicts the vector and raster views of some entities in the real world.

3.4 Vector Data Structure

Vector is one data structure that is widely used to store spatial data. The vector data model represents geographic features similar to the way maps do. It comprises lines or arcs, defined by start and end points, which meet at nodes. Vector data structures were the first to be used for digital cartography and GIS as they could simply be derived through digitization, are more exact in representing complex features and could be drawn on pen-type output devices like plotters. A vector-based GIS is defined by the vector representation of its geographic data. According to the structure of this data model, the locations of the nodes and the topological structure are usually stored explicitly. Features are defined by their boundaries only and curved lines are represented as a series of the connecting arcs. The explicit topology ensures that the structure only stores those points, which define a feature like point, line or arc. Everything outside these features is 'nonexistent'. The vector-based GIS (e.g., Arc/Info, ArcView) maintains double database—one for the spatial characteristics of the geographic objects (*where*?) and the other is the thematic data related to the characteristics associated with these objects (*what*?). Hence this is known as the *hybrid system* where the attribute data (relational table) is linked to the topological data through key element termed 'identifier', which is unique for each object.

3.4.1 Geometry and Topology

The *geometry* of a point is given by two-dimensional array while the line and area are given by a series of point coordinates, as shown in Fig. 3.3. Thus a point (a zero-dimensional abstraction of an object given by a single coordinate) is represented by a (x, y) coordinate, which implicates that the geographical extent of an object is limited to a location that can be specified by one set of (x, y) coordinates. A point normally represents a geographic feature too small to be displayed as a line or an area e.g., the location of a building or a point from where soil sample have been collected. A line is expressed by sets of (x, y) pairs i.e., a set of ordered coordinates that represent the shape of geographic features too narrow to be displayed as an area at a given scale (contours, roads or streams), or linear features with no area (country boundary lines). In Arc/Info, a line is synonymous with an arc. An area (polygon) is defined as a set of points representing the boundary. A polygon is defined by the lines that make up its boundary plus a point inside its boundary for identification. Polygons have attributes that describe the geographic feature they represent.

Depending on details one likes to reveal, a town can be represented by a point entity at some level of abstraction and a polygon entity at a higher level. If the level of resolution still increases, its internal structure may be revealed (e.g., the town can now be divided into zones, cluster of houses or even a single house and other details can be shown as shops, lamp posts etc).

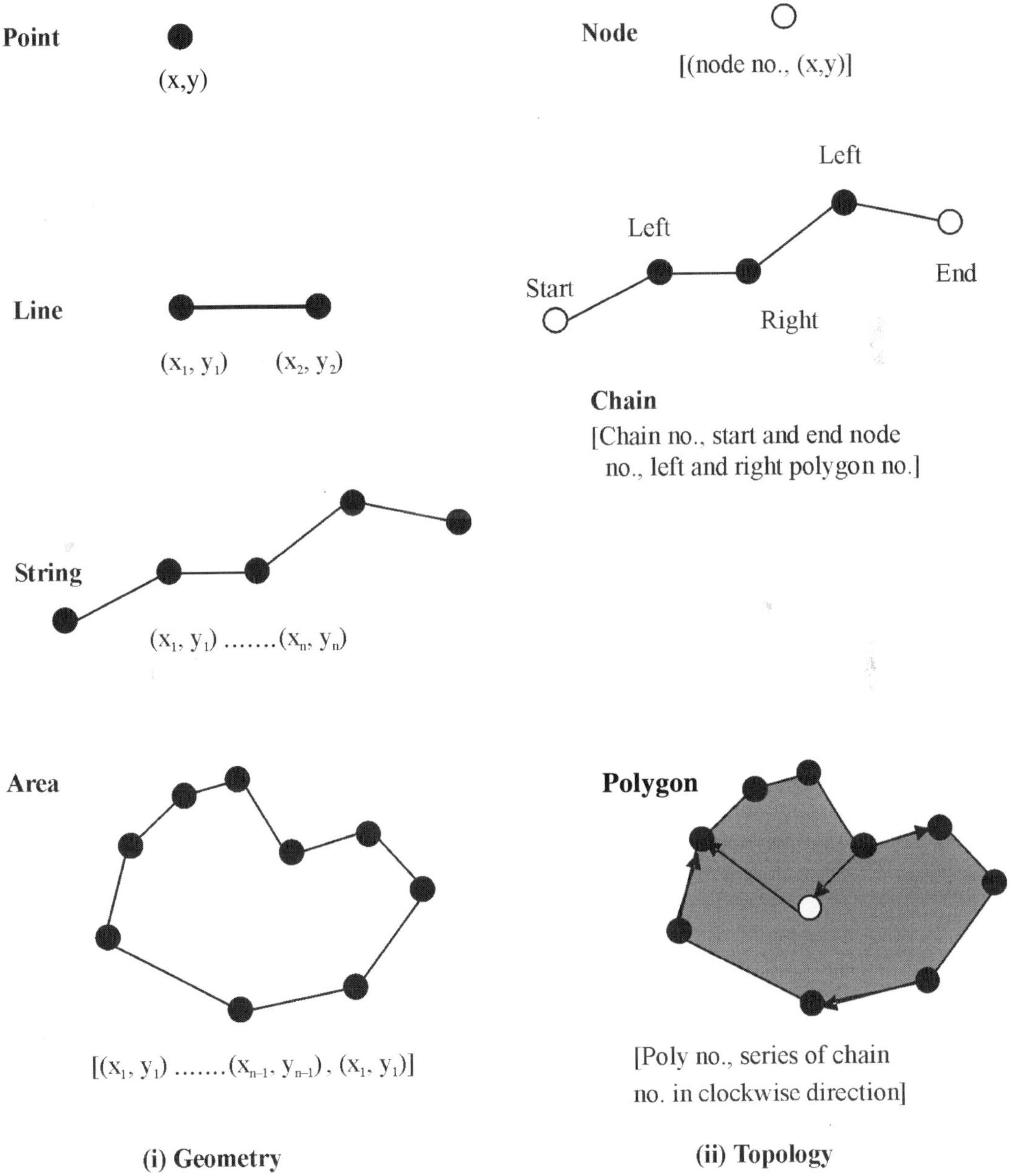

Figure 3.3: Geometry and topology of vector data

More complex definitions of point, line and area involve the functionality like the topological links between objects. For spatial analysis in GIS, only the geometry with the position, shape and size in a coordinate system is not enough but the *topology* is also required like how a line is linked to a polygon or a network. Topology refers to the relationships or connectivity between spatial objects. The topology of vector data is shown in Fig. 3.4. The topology is to avoid duplication in digitizing common boundaries of two polygons and to solve problems when the two versions of the common boundary do not coincide. It is absolutely necessary to build correct topological data sets without any single error to work with any GIS project.

Chain geometry		
Chain	Start node	End node
1	$[(x_{N1},y_{N1}),(x,y)]$	$[(x_{N2},y_{N2}),(x,y)]$
2	$[(x_{N2},y_{N2}),(x,y)]$	$[(x_{N3},y_{N3}),(x,y)]$
⋮		
6	$[(x_{N2},y_{N2}),(x,y)]$	$[(x_{N4},y_{N4}),(x,y)]$

Topology of polygon	
Polygon	Chain
100	1,2,5
101	–2,6,–3
102	4,–5,3

Topology of node	
Node	Chain
N1	1,–5,–4
N2	–1,2,6
N3	–2,3,5
N4	–3,4,–6

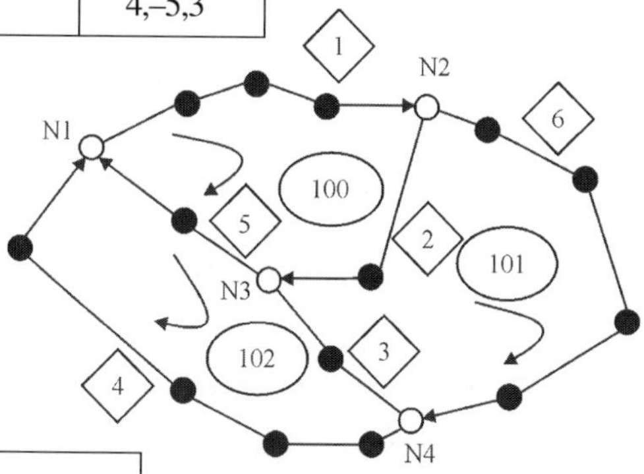

Topology of chain				
Chain	From	To	Left polygon	Right polygon
1	N1	N2	0	100
3	N3	N4	101	102

Figure 3.4: Building topology of vector data

3.4.2 Topology

In general, the notion, "topological" implies that certain relationships (connectivity, order, adjacency etc. are stored, making the data more useful for various kinds of spatial analysis. Let us explain it further in GIS, a line is also referred to as arc (so the name 'Arc/Info' or 'ArcView'). The arc stores with it the polygon at left or right, the forward and reverse arc linkages and the arc nodes for self-testing. This means that each line is stored only once (ruling out the redundancy in data—the only duplication is the end points). Topology allows this error correction, if any. One must see whether the first and the end points of an arc are the same to construct a polygon or whether a polygon is adjacent/just next to another polygon, or if there is any connectivity between two lines and so on. These relationships are well defined in a topological database, in contrast with the 'cartographic' database where these relationships are absent. Thus, cartographic databases are less useful for analysis of spatial data, however they are satisfactory for simple mapping. A cartographic database can usually be converted to a topological database by computing relationships—the process called "building topology".

One must remember that topology is concerned with the connectivity between entities and not their physical shape. Once the connectivity is made between a set of polygons with no gaps at nodes, the set is 'topologically clean'. Digitizing involves many errors, and it is the topology only that can give a check. A polygon gets its label from the label point (a point digitized inside the polygon). During digitization using any software, each end point is checked whether it is close to another in relation to the tolerance value (specified before). If it is so, the points are snapped together to make a single point. Again the arcs that connect the same nodes are examined to see if they are duplicates, and the user is asked to delete one of them. Any small areas resulted by double digitization of a single line (called *slivers*) thus get eliminated. The tolerance level, sometimes termed as *fuzzy tolerance*, has to be chosen very carefully, so that the separation between nodes can be treated separately. Short lines, small polygons or precisely measured point locations, if having significance for the map should not be deleted by automatic testing for topological completeness. Any map with features where topology has been built successfully, is a *complete* GIS map. Hence, topology is a must before any operation involving two or more maps. In short, topology ensures that:

1. No node or line segment is duplicated;
2. Line segments and nodes can be referenced to more than one polygon;
3. All polygons have unique identifiers; and
4. Islands and polygons with holes can be adequately represented.

To have a better understanding of topology, let us take an example of a blue print of a residential plot. The plot consists of four rooms with a small garden at the center. The vector data structure of this plot is presented in Figure 3.5. The structure may be built up in stages as described by Burrough (1998). The stages are: (i) linking arcs into a boundary network (the arcs which are topologically close to one another must be also together in the data file, thereby saving time and redundancy); (ii) checking polygons if they are closed or not; (iii) linking lines with polygons (the outer boundary of the map, called the 'envelope polygon,' must be built up and subsequently all the individual polygons to be created); (iv) computing polygon areas (following trapezoidal rule: area of a trapezium = ½ * sum of the sides * horizontal distance); and finally, (v) associating nongraphic attributes to the polygons to build the complete spatial database.

The *absolute location* of an object is an answer to a simple question "where", however, in modern geography, *relative location* is more important. Relative location reflects that spatial distances between objects are dynamic and the change in space depends on a number of factors like technological progress in transportation and communication, political situation and so on. Same is true for spatial events. In this context, the spatial relations between objects, as described by 'topology' altogether becomes more important.

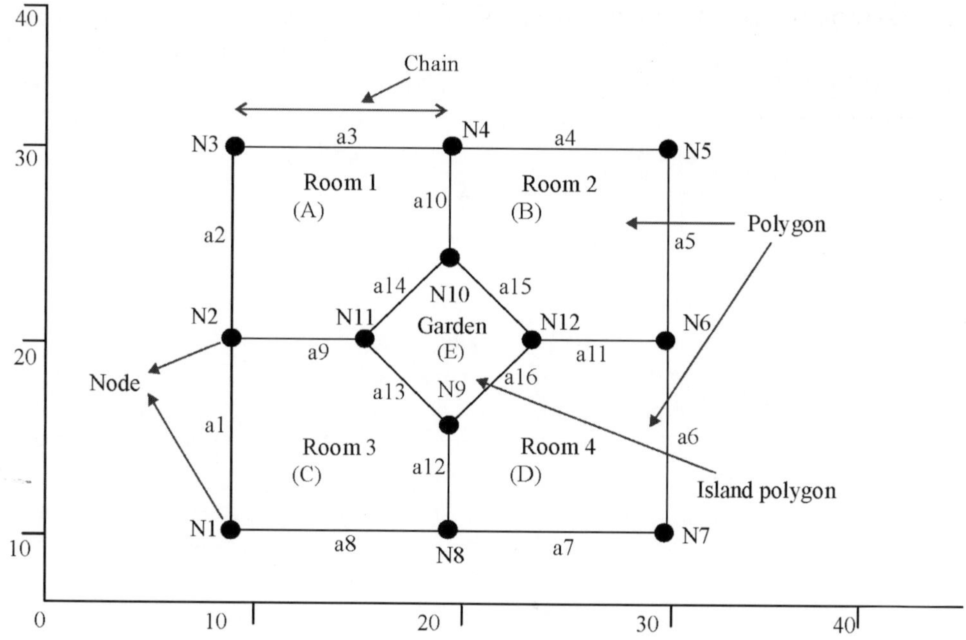

Node topology		
ID	X	Y
N1	0	10
N2	0	10
N3	0	10
N6	30	20
N9	20	13
N12	23	20

Polygon topology	
ID	Vector
A	a1, a9, a13, a12, a8
B	a2, a3, a10, a14, a9
C	a4, a5, a11, a15, a10
D	a6, a7, a12, a16, a11
E	a13, a14, a15, a16

Vector topology				
ID	Start-node	End-node	Left-polygon	Right polygon
a1	N1	N2	Outside	A
a2	N2	N3	Outside	B
a3	N3	N4	Outside	C
a8	N8	N1	Outside	C
a16	N12	N9	D	E

Figure 3.5: Complex topology of vector data

3.4.3 Topological relationships between spatial objects

There are a vast number of possible relationships in spatial data of which many are important for analysis in GIS. For example, "is contained in" relationship between a point and an area is important in relating objects to their surrounding environment. Likewise, "intersects" between two lines is important in analyzing routes through networks. Relationships can exist between entities of the same type or of different types. For each shopping center, we can find the nearest shopping center (same type); again, for each customer, we can find the nearest shopping center (different types). In practical applications of GIS, all possible relationships in spatial data should be used logically with more complicated data structures. There are three types of relationship as given below.

1. Relationships that are used to construct complex objects from simple primitives such as; a) relationship between a line (chain) and the ordered set of points which defines it; b) relationship between an area (polygon) and the ordered set of lines which defines it.
2. Relationships that can be computed from the coordinates of the objects, such as; a) two lines can be examined to see if they cross ("crosses" relationship); b) areas can be examined to see which one encloses a given point ("is contained in" relationship); c) areas can be examined to see if they overlap ("overlaps" relationship).
3. Relationships that cannot be computed from coordinates and must be coded in the database during input such as (a) we can compute if two lines cross, but not if the highways they represent intersect (may be an overpass); (b) some database allows an entity called a complex object, composed of simple objects, e.g., objects representing house, lot, cabel, with associated attributes grouped together logically as "account".

Considering the basic entities, point, line and polygon, the following topological relationships between them do commonly exist (Fig. 3.6).

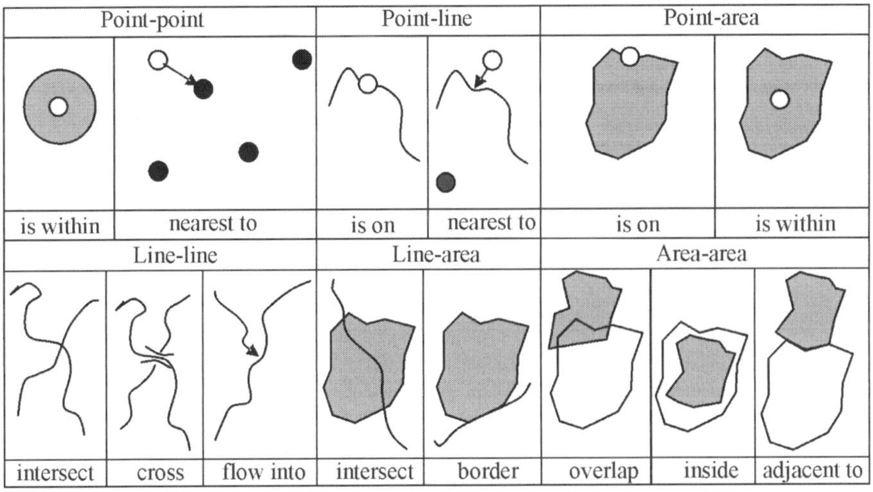

Figure 3.6: Topological relationships between spatial objects

Point-point

"is within": within a certain distance (the customer within 1 km of a retail store)

"is nearest to": nearest to a certain point (the hazardous waste site nearest to the groundwater well)

Point-line

"on line": a point on a line (a boat on the river)

"is nearest to": a point nearest to a line (the road nearest to the construction site)

Point-area

"is contained in": a point in an area (the district headquarter in the district)

"on border of area": a point on the border of an area (a tubewell placed at the border of a ground)

Line-line

"intersects": two lines intersect (two roads intersect at a T-point)

"crosses": two lines cross without an intersect (the road crosses a river through a bridge)

"flow into": a drain flows into a river

Line-area

"intersects": a line intersects an area (the soil types crossed by the railroad)

"borders": a line is a part of border of an area (a road forms part of the boundary of a district)

Area-area

"overlaps": two areas overlap (between soil types and land use of an area)

"is within": an island within an area (a lake in a garden)

"is adjacent to": two areas share a common boundary (two neighboring adjacent districts)

The geometric and topological modelling between point, line and area is presented in Figure. 3.7.

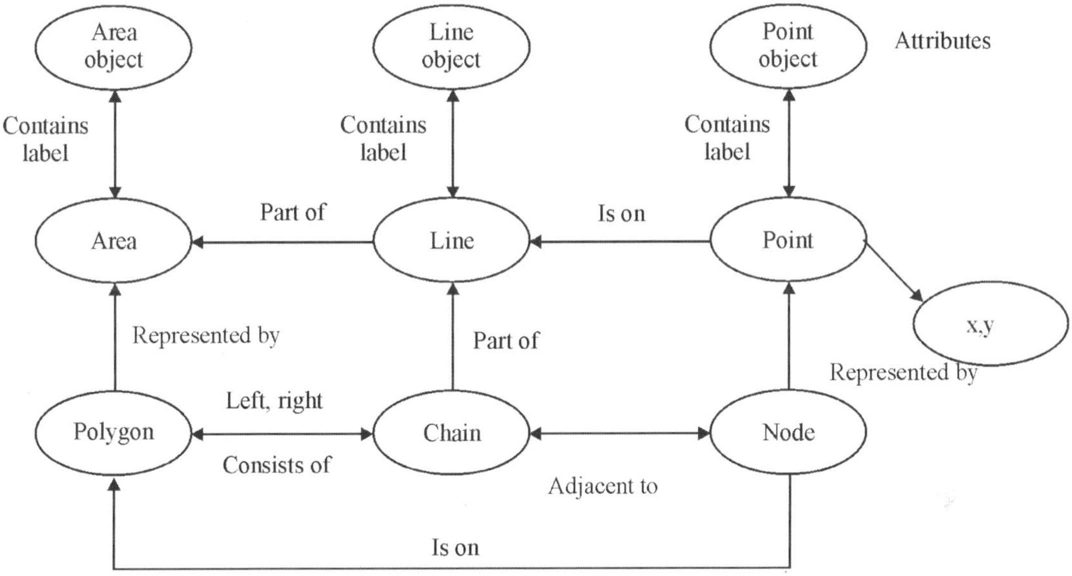

Figure 3.7: Geometric and topological modeling

3.4.4 Vector Models

As explained earlier, the vector data model is a relatively efficient means of storing the geometric information of spatial data but with difficulty in accessing the topological and attribute data. There are a variety of vector models to store and manage vector information (Lakhan, 1996; Bernhardsen, 1999; DeMers, 2000). Each of them are explained here with their advantages and limitations.

1. **Spaghetti model** This is the simplest vector model with one-to-one translation of the graphical image on the map. As the name (given by Nick Crisman, a programmer) suggests, these earliest vector files are simple lines with arbitrary starting and ending points just like following the path of a single stand of spaghetti through a pile of spaghetti on one's plate. Thus the files written in binary or ASCII code contained some long and some short lines or a mix of the two. This unstructured vector data format is called 'cartographic spaghetti model' (Fig. 3.8a). A number of systems still use this structural format, but most GIS systems allow to transfer data in this format and subsequently convert them into topological data format. Though this data model is simple and easy to manage, it contains no topology and involves lots of duplication rendering more space for data storage. An advancement of spaghetti model is the *vertex dictionary model* where only the geometric information of the vertices are stored once in a file and another file is maintained for the identification of features, thus eliminating duplicity in storing data (Fig. 3.8b).

2. **Arc/node model** This model has the same structure as in case of hierarchical system of organizing attribute database. As area consists of connected lines, and the line of connected points, each feature (point, line or polygon) can be separated into another with its immediate earlier dimension. Thus a line consists of one arc and a polygon of more than one arc, each of them with a set of points. In this data model, we have at least one file containing the attributes of polygon, another file containing the list of arcs and third file with the coordinates of the points that constitute the arcs. Each file has a link with the immediate higher order file as shown in Fig. 3.9. Many first generation systems (Polyvet, Giras and Odyssey) used to follow this data model.

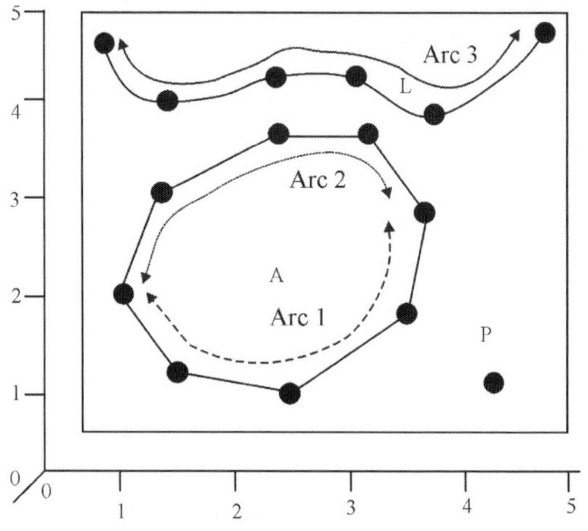

A, 8 (identifier of polygon; number of vertex)
1,2 (coordinate of first vertex)
1.3,1.2
2.4, 1
3.4, 1.8
3.5, 2.7
3, 3.5
2.4, 3.5
1.4, 3
1,2 (coordinate of the first vertex)

L, 6 (identifier of line; number of vertex
0.9,4.8
1.3, 4
2.5, 4.2
3, 4.1
3.8, 3.9
4.9, 4.8

P, 1 (identifier of line; no. of vertex)
4.2, 1

(a)

File 1

Vertex	*X*	*Y*
1	1	2 (coordinate of first vertex)
2	1.3	1.2
3	2.4	1
4	3.4	1.8
5	3.5	2.7
6	3	3.5
7	2.4	3.5
8	1.4	3
9	0.9	4.8
10	1.3	4
11	2.5	4.2
12	3	4.1
13	3.8	3.9
14	4.9	4.8
15	4.2	1

File 2
Polygon A: 1,2,2,4,5,6,7,8
Line L: 9,10,11,12,13,14
Point P: 15

(b)

Figure 3.8: Vector data model (a) spaghetti model and (b) vertex dictionary

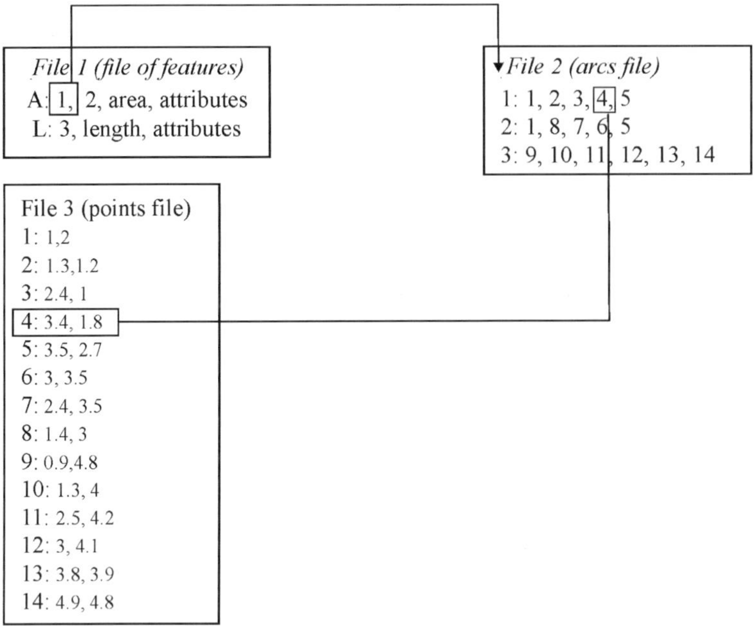

Figure 3.9: Arc/node vector data model

3. **Topological model** With the first International Advanced Study Symposium on Topological Data Structures for GIS (1979), an improved mode of vector data model has been introduced. This model allows easy performation of many advanced analytical techniques and thus, incorporates some of the more often used operations in advanced GIS analytical techniques (DeMers, 2000; Dangermond 1982). This data model (Fig. 3.10) excludes the file storing attributes of polygon as a separate file, rather it considers arc as the basic unit for data storage; thus the information of adjacency etc., are stored explicitly into the data model. The point data are associated with the arc file as only the first and the last points of the arc called *start* and the *end nodes*. Similarly the polygons are also associated as *left* and *right polygon* of the arc considered. During the digitizing process, one arc needs to be digitized once, thereby leaving thechance of duplicity while tracing adjacent polygons connected by a common arc. This completely eliminates *sliver* (a type of error) during digitization.

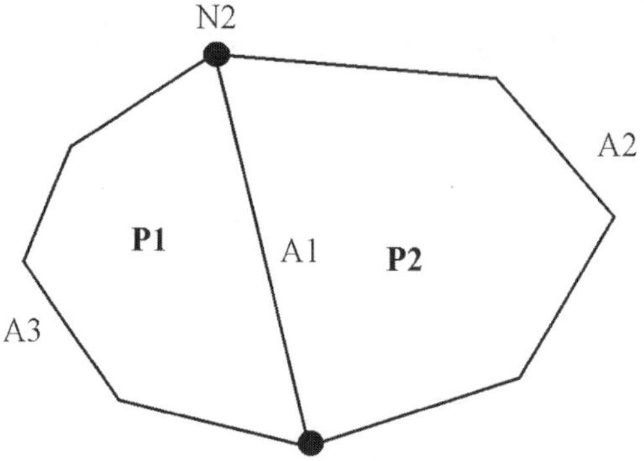

Arc file

Arc	From node	To node	Left poly	Right poly	From node(x,y)	To node(x,y)
A1	N1	N2	P1	P2	N1x, N1y	N2x, N2y
A2	N1	N2	P2	outside	N1x, N1y	N2x, N2y
A3	N2	N1	outside	P1	N2x, N2y	N1x, N1y

Figure 3.10: Simplified scheme of topological vector data model

4. **Triangulated Irregular Network (TIN)** Vector data models, unlike raster models, do not perform well with surface like topography or land use or air temperature. In order to fit vector model to work with geographic structure, a team of geographers and computer scientists came out in early 1970 with a new data structure called Triangulated Irregular Network (TIN), which is an alternative to the raster model of a DEM. This was a significant achievement which has later (in 1980s) been adopted in many GIS and automated mapping and contouring packages.

The Triangulated Irregular Network model is a simple way to build a surface from a set of irregularly spaced points with their coordinates. We know that any two points make a line and three make a plane. In a TIN, the irregularly spaced points are connected by lines to form triangles under Delaunay criterion (more detailed discussion in Chapter 5) forming a mosaic of surface (Fig. 3.11). Each triangle represents a piece of the surface (plane) and accurately fit into the neighboring piece to make the surface continuous as a set of contiguous, non-overlapping triangles (as each piece of the mosaic surface is defined by the three vertices of the triangle). Thus, TIN is viewed as polygons having attributes of slope, aspect and area, with the vertices of triangle as elevation attributes and the three edges of triangle as slope and direction attributes. Hence the points are stored in a file containing information regarding the topology of a network.

TIN model for a surface with a rough terrain is made up of more points and for a smooth surface it requires few points. If the sample points representing the surface are carefully selected, the model will be very accurate. Well-placed sample points occur where there is a major change in the shape of the surface, for example, at the peak of a mountain, the floor of a valley, or at the edge (top and bottom) of cliffs. TIN is very effective for certain types of terrain, which can be very effectively divided into triangles with plane facets, but for other landscapes, such as glaciated ones, TINs are not effective as the surface cannot be well represented by flat triangles. The advantages of TIN over GRID (a raster data model to represent surface) are its ability to describe surface at different levels of resolution to show areas with different complexity of relief and its efficiency in storing data. But in many cases, TIN requires visual inspection and manual control of the network and difficult to integrate with raster databases (such as satellite imagery).

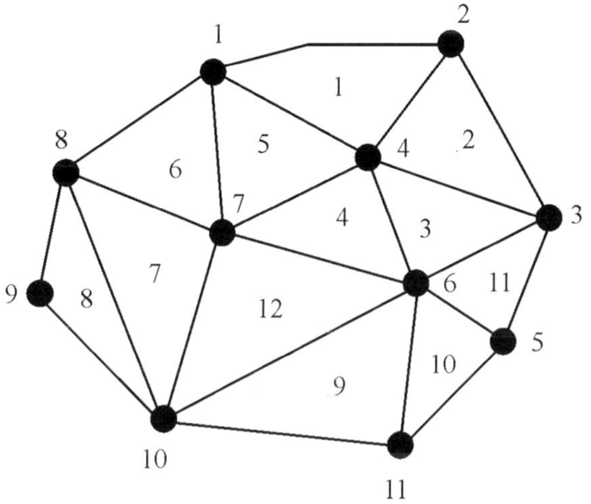

Node file			
Node	X	Y	Z
1	X1	Y1	Z1
2	X2	Y2	Z2
3	X3	Y3	Z3
4	X4	Y4	Z4
5	X5	Y5	Z5
6	X6	Y6	Z6
7	X7	Y7	Z7
8	X8	Y8	Z8
9	X9	Y9	Z9
10	X10	Y10	Z10
11	X11	Y11	Z11

Triangle	Vertex	Neighbor triangle
1	1,2,4	Outside, 2, 5
2	2.3.4	3, 1, outside,
3	3,4,6	11,4,2
4	4,6,7	3,12,5
5	1,4,7	1,6, outside
6	1,7,8	5,7, outside
7	8,7,10	12,8,6
8	8,9,10	7, outside, outside
9	6.11,10	10, outside, 12
10	6,5,11	11, outside, 9
11	3,5,6	3, outside, 10
12	7,6,10	4,9,7

Figure 3.11: Triangulated irregular network

3.4.5 Vector data format

Most GIS can import different data formats, or use utility programs to convert them. Data formats can be industry standard or any format that is commonly accepted. Vector formats are either page definition languages or preserve ground coordinates. Common page languages are

Postscript, HPGL or AutoCAD DXF. True vector GIS formats are Digital Line Graph (DLG) and Topologically Integrated Geographic Encoding and Referencing (TIGER) that have topology. The data structures are specific and internal to any GIS and generally not exposed to the user. Any GIS needs to import data from other GIS packages or from the most common data sources, or through scanning or digitization process. The imported data is then converted to internal format specific to the GIS system being used. This internal or exchange data formats are open for some system (Intergraph / Autodesk) but a trade secret for other (ArcView, IDRISI, ArcGIS). Some industry used standard vector data formats are HPGL (Hewlett-Packard Graphics Language), Adobe Corporation's Postscript page definition language and Autodesk's AutoCAD Digital Exchange Format (AutoCAD DXF). DXF is probably the most widely used vector data transfer format, and a file in DXF format offers some very strong advantages. It contains a complete display information, and almost every graphic program can read it. However, there are several different ways to store attribute information in DXF and to link DXF entities to external attributes. Because, there are no attribute standards, many programs that claim to read DXF files still do not import attribute information properly. HPGL is a language that controls computer plotters. It contains display information but no geographic coordinates or attribute data. It is usually not appropriate for the storage or transfer of GIS data.

None of these formats support true topology, but can maintain information in separate layers—a very basic GIS functionality. Two most commonly used vector data formats (used since long) are US Geological Survey's Digital Line Graph (DLG) and US Census Bureau's Dual Independent Map Encoding (DIME). These formats contain topology and have been in use extensively. DLG is a transfer format used by the US Geological Survey (USGS), depicts vector information portrayed on printed-paper maps. It carries very accurate coordinate information and sophisticated, feature-classification information but no other attribute data. DLG does not include any display information. The DLG standard is significant because the USGS and other US government agencies have used it to publish large number of digital maps.

Topologically Integrated Geographic Encoding and Referencing (TIGER) is an ASCII transfer format used by the US Census Bureau to store the street maps constructed for the 1990 census. It contains complete geographic coordinates and is line, not polygon, based (although polygons can be constructed from its attribute information). The most important attributes include street name and address information. TIGER does not contain display information. Maps of the entire US are available in TIGER format.

A new transfer format was developed by the US government called *Spatial Data Transfer System (SDTS)* to handle all types of geographic data. SDTS can be either binary or ASCII but is generally binary. Virtually all geographic concepts can be encoded in SDTS, including coordinate information, complex attribute information, and display information. This versatility causes a corresponding increase in complexity. To simplify things, several standard subsets of SDTS have been adopted. The first of these, the *Topological Vector Profile* (TVP), is used to store certain types of vector maps. SDTS can also be used for raster information. Not much data is available in SDTS format at this time, nor do many software systems support it. However, it will be the foundation of the US National Spatial Data Infrastructure (NSDI). Its importance will increase as more NSDI data becomes available.

3.5 Raster Data Structure

A continuous surface can be discretized into sets of single basic units, such as square, triangular, or hexagonal cells, or into irregular triangles or polygons (Thiessen polygons) which are tessellated to form geographical representations. Use of irregular triangles is based on the principles of triangulation (continuous surface of land is approximated by a mesh of triangles whose apices or nodes are given by some measured parameters). These may be used in continuous variation in the dynamic modeling of surface water movement, groundwater flow, wind movement etc. An alternate to this triangulation is the regular tessellation or regular grid. The two-dimensional surface is divided into square cells (known as *pixels*) whose size is determined by the resolution that is required to represent the variation of an attribute for a given purpose (Fig. 3.12). The resolution may vary from centimeters to many kilometers. This grid cell representation of space is known as *raster data model*. Here the area is divided into rows and columns, which form a regular grid structure. Each cell must be rectangular in shape, but not necessarily square. Each cell within this matrix contains location coordinates (x, y) and a discrete attribute value attached to it. These attribute data are two-dimensional to store various information such as forest cover, soil type, land use, wetland habitat, or other data in different layers (layers are functionally related map features). Raster requires less processing than vector data, but it consumes more storage space in computer since they store each cell in the matrix regardless of whether it is a feature or simply 'empty' space. Scanned maps or remotely sensed data (satellite imagery) are in raster format. Digital terrain models (DTM) and digital elevation models (DEM) are two most common examples of raster data.

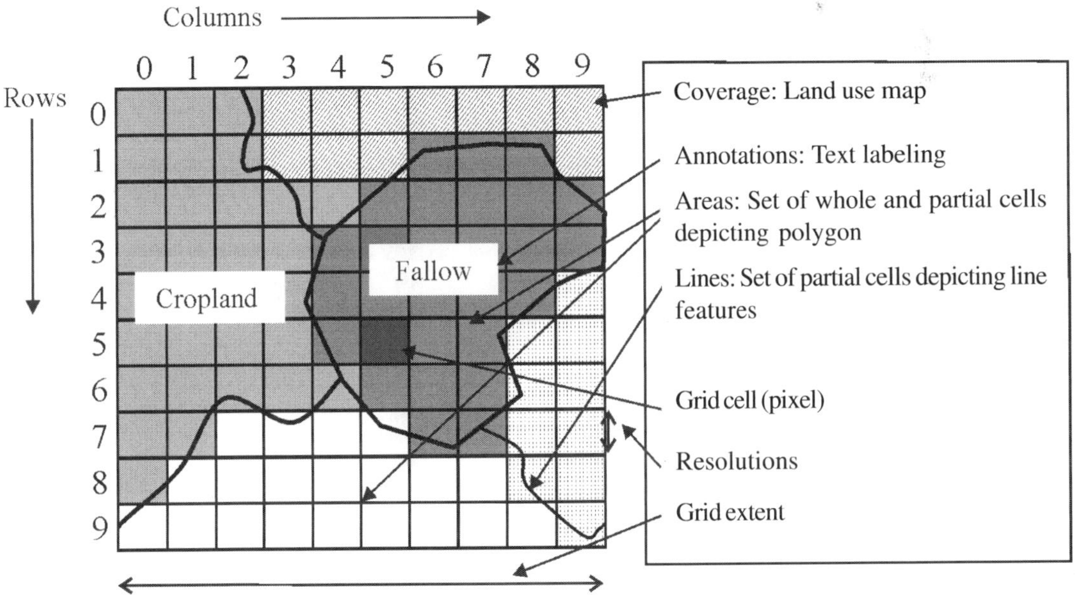

Figure 3.12: A raster map

There are two views on the meaning of a pixel in a raster model. First, the pixel can be seen to represent a singular grid point. In this case, the rectangular area enclosed by four neighboring grid points is called a *facet* of the raster model. From another point of view, the pixel can be seen to represent a rectangular area itself in an integral manner. The value assigned to a pixel describes one thematic attribute of that pixel. Topological information is only contained 'implicitly' within the ordering of the matrix (in a vector structure topology is stored 'explicitly'); neighborhood relations can be described by index differences. Topographic objects can only be represented by a set of neighboring pixels having identical attributes. Thus, the manipulation of individual objects becomes very difficult. The location of each cell calculated from the origin of grid: "*two down, three over*" (usually from upper left, but lower left in ArcView).

→ Columns

Rows ↓	0	1	2	3	4	5	6	7	8	9
0	M	M	M	M	M	V	V	V	F	F
1	M	M	M	M	M	V	V	V	F	F
2	M	M	M	M	M	V	V	V	F	F
3	M	M	M	W	W	W	W	F	F	F
4	M	W	M	W	W	W	W	F	F	F
5	W	W	W	W	W	W	W	F	F	F
6	W	W	W	W	W	W	W	F	F	F
7	W	W	P	P	P	W	2	F	F	F
8	W	W	P	P	P	W	2	F	F	F
9	W	W	W	W	W	W	F	F	F	F

Pixel	Value
0	M
1	M
2	M
3	M
4	M
5	V
6	V
7	V
8	F
9	F
10	M

Pixel	Value
42	W
43	W
44	F
45	F
46	F

Pixel	Value
76	W
77	W
78	W
79	F
80	F
81	F

Figure 3.13: Exhaustive representation of raster

Looking closer, raster data is an abstraction of the real world where spatial data is expressed as a matrix of cells or pixels, with spatial position implicit in the ordering of the pixels (Fig. 3.13). With the raster data model, spatial data is not continuous but divided into discrete units. This makes raster data particularly suitable for certain types of spatial operation, e.g., overlay operation or area calculations.

In raster format, individual cells are used as the building blocks for creating images of point, line, area, network and surface entities. The geometry of raster data is presented in Fig. 3.14. A point is given by point ID, coordinates (i, j) and the attributes; a line is given by line ID, series of

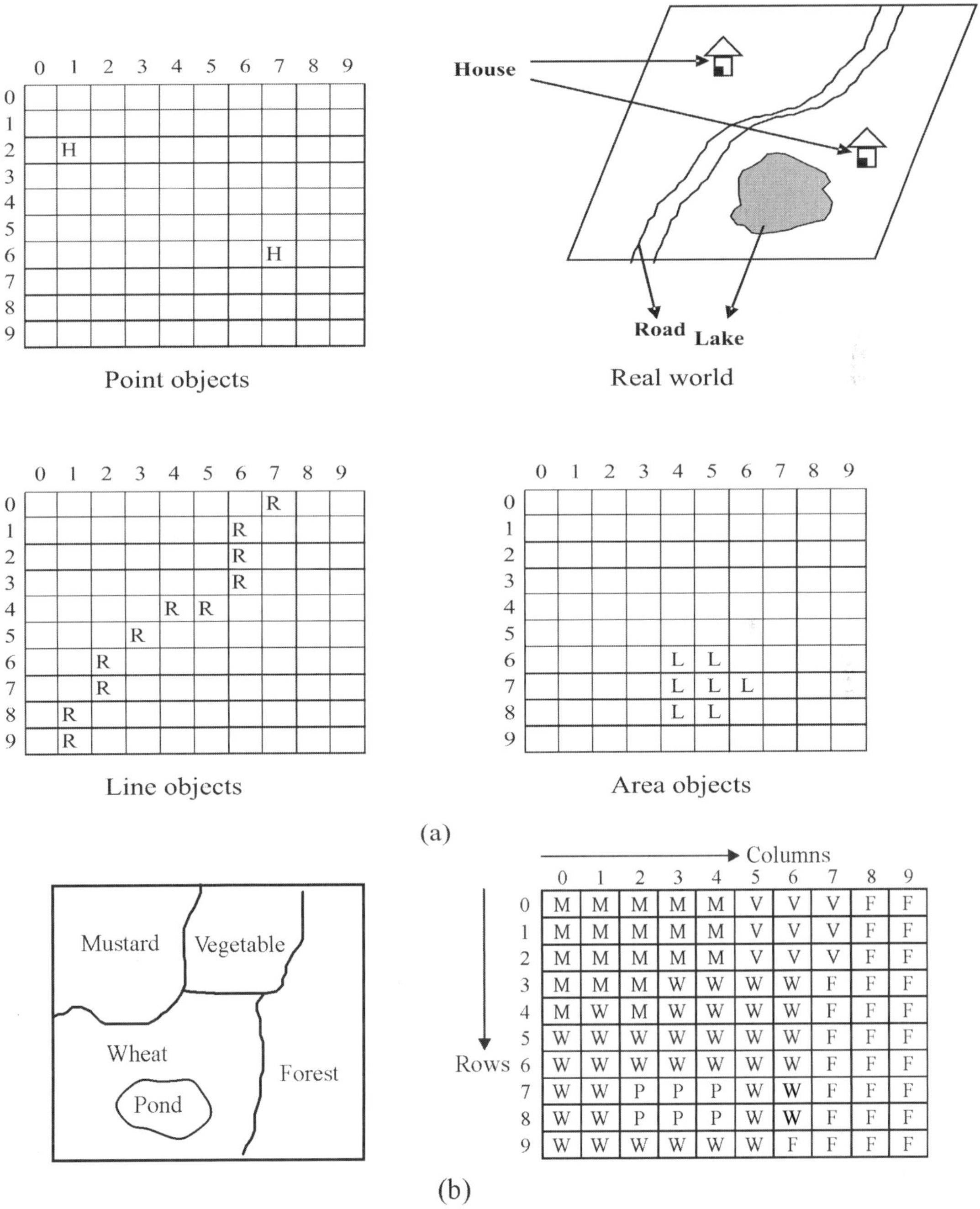

Figure 3.14: (a) Geometry and topology of raster data (b) a raster data model with multiple land use

coordinates forming the line, and the attributes; and an area segment is given by area ID, a group of coordinates forming the area and the attributes. The topology of line objects is given by a sequence of pixels forming the line segments. Area objects in raster model are typically given by "run length" that rearranges the raster into the sequence of length (or number of pixels) of each class, "run length" structure (explained later in details) is as follows:

- start line number (start pixel number, number of pixels);
- second line number (start pixel number, number of pixels);

The topology of raster model is relatively simple as compared with the vector model. One of the weak points in raster model is the difficulty in network and spatial analysis as compared to the vector model. For example, though a line is easily identified as a group of pixels which form the line, the sequence of connecting pixels as a chain would be a little difficult in tracing. In case of polygons in raster model, each polygon is easily identified but the boundary and the node (when at least more than three polygons intersect) should be traced or detected.

3.5.1 Creating a Raster

Let us consider laying a grid over a geologic map. Coding each cell with a value that represents the rock type for the majority of that cells areas makes a raster and every cell in it must have a coded value. In most cases the values that are to be assigned to each cell in the raster are written into a file called ASCII files (see Box 3.3). This file can be created manually by using a word processor, database or spreadsheet program or it can be created automatically. It is then normally imported into the GIS so that the program can reformat the data for its specific processing needs. There are several methods for creating raster database of which the two most important methods are given below (Start & Estes, 1990).

1. Cell by Cell Entry Direct entry of each layer through cell by cell is the simplest. Entry may be done within the GIS or into an ASCII file for import. The process is normally tedious and time-consuming as a layer may contain millions of cells (e.g., average Landsat image consists of around 7.4 × 106 pixels and an average Thematic Mapper (TM) scene is made of about 34.9 × 106 pixels).

2. Digital Data This method is useful and preferred for image data like satellite remote sensing imagery. However, resampling will likely be needed in order that pixels coincide in each layer.

One must remember that each cell or pixel has to be assumed only one value. The type of values contained in cells depend upon both the reality being coded and the GIS. Different systems allow different classes of values, including whole numbers (integers), real (decimal) values, or alphanumeric values. Many systems only allow integers, others that allow different types restrict each separate raster layer to a single kind of value. It is unreasonable to try to multiply the values in a numeric layer with the values in a nonnumeric layer. Integer values often act as code numbers, which "point" to names in an associated table or legend, e.g., the integers correspond to the legend identifying the name of each soil class may be 0 = "no class", 1 = "fine sandy loam", 2 = "coarse sand", 3 = "gravel" etc. The data for an area can be visualized as a set of maps of layers. A map layer is a set of data describing a single characteristic for each location within a bounded geographic area. Only one item of information is available for each location within a single layer. Multiple items of information require multiple layers. A topographic map can show multiple items of information for each location like elevation (contours), districts (boundaries), roads, railroads, urban areas etc. There would be *five* layers in a raster GIS. A raster database may contain up to a hundred layers and each layer typically contains hundreds or thousands of cells. The important characteristics of a layer will be its resolution, orientation and zone(s).

Resolution In general, resolution can be defined as the minimum linear dimension of the smallest unit of geographic space for which data are recorded. In the raster model the smallest units are generally rectangular (occasionally hexagons or triangles). These smallest units are known as cells or pixels (high resolution refers to raster with small cell dimensions i.e., more details).

Orientation The angle between true north and the direction defined by the columns of the raster.

Zones Each zone of a map layer is a set of contiguous locations that exhibit the same value (like ownership plots, political units such as counties or nations, lakes or islands, individual patches of the same soil or vegetation type). The term *class* is often used to refer *zones*. It is to be noted that all map layers will not have zones; cell contents may vary continuously over the region making every cell's value unique (e.g., satellite sensors record a separate value for reflection from each cell). Major components of a zone are its value (information stored in a layer for each pixel or cell) and location(s) (identified by an ordered pair of coordinates—row and column numbers).

The major problem with raster data is the *mixed pixel* problem. Consider Fig. 3.12. Here, the fallow land and cropland are adjacent. The pixels at the common boundary of the two features (*edge pixels*) belong to either of the two categories, cropland or fallow. Thus, it is hard to assign these pixels to one or other category. A compromise has to be made to assign these edge pixels in one class or another. Hence, we either have to make up rules for assignment, such as assigning a mixed pixel to the class that occupies the most area within it, and so on.

3.5.2 Raster data format

Raster data formats have been used since digital images came into existence. These formats are easy to pass over through networks and are very simple to create. The most common raster data formats are:

1. TIFF (Tagged Image File Format)

Developed by Aldus and Microsoft Corporation and primarily designed for raster data interchange, TIFF format is highly flexible and platform-independent format which is supported by numerous image processing applications including remote sensing and multispectral data. TIFF is also able to decompose an image by tiles rather than scan-lines, thereby permitting a very efficient access to a large compressed imagery. A number of TIFF extensions are available for specific applications like GeoTIFF, Adobe extensions for Page Maker, JPEG-in-TIFF, Zip-in-TIFF etc.

2. GIF (Graphic Interchange Format)

Created by CompuServe, this format defines a protocol intended for the on-line transmission and interchange of high-quality, high-resolution raster graphic data in a way that is independent of the hardware (i.e., machine independent) used in their creation or display. GIF files can contain one or more images with 256 colors or less. The images are compressed using LZW compression, reducing the file size by 30-50 per cent or more. It is flexible and has capabilities that JPEG does not, such as animation and transparency.

3. Joint Photographic Experts Group (JPEG)

In the early days of the web, browsers (a web browser is a software that translates **H**yper **T**ext **M**arkup **L**anguage or HTML files into **W**orld **W**ide **W**eb or WWW documents that integrate text and images) supported only the GIF format images, until the JPEG format (also JPG), a format which compresses the image file so that it takes up less disk space, was introduced. JPEG allows images to be displayed in millions of colors, while GIF supports only 256. The GIF format is not as well suited to compressing pictures with semi-random patterns of dots of differing colors, and it is even less suited to compressing images, which have extremely subtle gradations of colors such as one might see in a photograph. On the other hand, the JPEG compression scheme has been designed to process photographic images. Rather than looking for patterns or runs of colors on a line-by-line basis, JPEG uses a mathematical formula based on the way our eyes perceive color. The JPEG compression algorithm looks at a pixel and its neighbors in all directions and finds the factors for that formula which will best represent all those pixels. Rather than storing the pixels, it compresses the image by storing only the factors to save space. When you view the pictures, the JPEG process plugs the factors back into the formula and generates the pixels that best represent the original image. However, some special effects like transparency and animation are not supported here. With the wide introduction of scanners, digital cameras and the Worldwide Web, the JPEG image format has quickly become the most widely used digital image format.

Some other raster formats commonly used are *Arc Digitized Raster Graphics* (*ADRG*) which is used by the US military to store raster images of paper maps; *Band Interleaved by Line* (*BIL*), *Band Interleaved by Pixel* (*BIP*), *and Band Sequential* (*BSQ*)—these are formats produced by remote-sensing systems like *Digital Elevation Model* (*DEM*): used by the USGS to record elevation of points on the Earth surface; *PC Paintbrush Exchange* (*PCX*)*:* a common raster format produced by most scanners and PC drawing programs; *Spatial Data Transfer Standard* (*SDTS*): a general-purpose format designed to transfer geographic information etc.

3.6 THE RASTER-VECTOR DATABASE DEBATE

Which among raster or vector is better—have been the argument in place since long. The appeal of raster database as its simplicity of organization and speed of many operations (like overlay, buffering) is immense, especially to the remote sensing community who are used to "pixel" processing. On the other hand, there are many situations in which the raster approach may appear to sacrifice too many details and also the "inaccuracy" caused by the cells when portraying linear features and points. However, computing times for overlaying vector-based information can be excessive. There are situations under which vector approach is more functional to raster or vice versa. Today, the question has evolved from "Which is best?" to "Under what conditions which one is the best and how can we have flexibility to use the most appropriate approaches on a case by case basis?" There are four issues to the discussions of raster *vs.* vector and these are coordinate precision, speed of analytical processing, mass storage requirements and characteristics of phenomena (Burrough, 1986; Gahegan & Roberts, 1988; Star & Estes, 1990).

3.6.1 Coordinate precision

In most cases it is not clear whether the center of the cell in a raster or one of its corners is the precise location of the coordinate. On the other hand, for certain classes of data like data captured from precision survey like soil mapping, maps created from land surveyors' coordinates (like survey of India maps), political boundaries defined by accurate survey, vector proves to be more realistic. However, true comparison in terms of precision should be between raster cell size and the positional uncertainty of a vector object, not the coordinate precision.

3.6.2 Speed of processing

Raster data can be processed very quickly to answer most analytical questions involving overlays, proximity, and Boolean queries. In most cases, raster analysis requires cell-by-cell comparison of the contents of layers and little arithmetic computation is required. Thus raster systems are suitable for small computers and have low program development costs. The same analysis require considerable computation in vector systems. Since vector topology is complex, so to solve complex geometrical problems (e.g., to find line intersection or polygon intersection points), complex geometrical algorithms are needed. These slow down the response time of vector based systems, preventing large scale work, or multiple users on small computers.

3.6.3 Mass storage

Simplest raster data storage method requires one memory location e.g., one or two bytes per cell and file compression is possible through a variety of approaches such as 'run length encoding'. Vector storage also uses very little storage for simple polygons, but its memory requirements depend on complexity of objects and the precision of coordinates (i.e., single or double). Generally, vector systems use less mass storage than a raster based system.

3.6.4 Characteristics of phenomena

Raster is a regularly spaced sampling of phenomena and it reflects lack of knowledge of spatial variation. But it allows sampling more heavily in areas of complex variation—not wasting samples in areas of little variation. Raster is appropriate for remote sensing as the satellite is not intelligent enough to vary its sampling in response to variation on the Earth's surface. Vector representation permits more spatial variability in some areas than in others. This is appropriate for social, economic, demographic variation, which is more intense in some areas than in others. It is also appropriate for some natural phenomena, which is dynamic (variation in vegetation cover is more rapid near the Ganges than in the arid tracts of Rajasthan).

Thus, we require systems, which act in two ways:

1. The system can store data in one form and process it in another through an efficient algorithm to convert from raster to vector and vice versa. It should be possible to capture and store data in vector mode, yet analyze it in raster form. This may save computing time and improve mass storage which are especially important for small machines.
2. Combinations of systems which run raster and vector systems in parallel, e.g., raster and vector systems on the same PC, use conversion functions in one or both systems or overlay a vector based land use parcel map over a satellite image to improve the interpretation of the image. This image might then be used to correct a vector based vegetation parcel map. Some basic differences between raster and vector data format and their relative advantages and limitations (UNESCO, 1999) are presented in Tables 3.1 and 3.2.

3.7 Translations Between Vector and Raster

Conversion between raster and vector data is essential for practical applications of GIS. In many cases transfers between the two main data models are necessary for the reasons of data display or printing. The most obvious example is the display (or printing) of vector data on a standard PC screen (or many types of printers). Most computer screens (or printers) apply raster technology, so any coverage must be transferred to raster before it can be properly displayed (or printed).

Transferring data format is also necessary from the methodological and analytical points of views. For example, translation is needed to analyze transportation routes in a vector system while area of study is derived from satellite imagery (raster format). In another case for monitoring land use/land cover of the same area, one must have the taluk/district/village boundaries (derived as vector) to start with. These boundaries, must be transferred in order to prepare it for pixel-based analysis.

Table 3.1 Raster *vs.* vector data structure

Characteristics	Vector data structure	Raster data structure
Location	Referenced by x, y coordinates that can be linked to form lines and polygons.	Referenced by a grid cell in a rectangular array of cells (matrix).
Attributes	Referenced through single value for that cell.	Represented as a unique ID number to tables.
Precision	Allows precise representation of points, boundaries, and linear features.	Precision is less.
Spatial analysis	Operations such as geometrical transformations or visualization can be quickly performed. But in some situations, e.g., in the context of the union of thematic layers, more complex algorithm have to be applied than with raster data defined on a common grid. Thus, overlaying multiple vector maps is time consuming and needs considerable computing power.	Spatial analysis is mostly easy and quick. Operations requiring information on surface coverage can be performed rather easily.
Simulation and modelling	Simulation of processes of special interaction is very difficult as topology is explained explicitly and each spatial entity has its own form and shape.	Spatial modelling is easy and fast as all spatial entities have a simple and regular shape.
Topology	Topology can be described explicity, so network analysis is better performed.	Implicit description of topology, hence topological associations are difficult to define.
Data structure	The structure of vector data is simple and economical in terms of data files.	Complex data structure and hence, takes more space in storing data.
Coordinate transformation	Easy and also rubber sheeting.	Difficult and time consuming unless special algorithms are used.
Graphic Applications	Accurate at all scales. Best for features with discrete boundaries and analysis which involves accurate positioning like political boundaries, roads, rivers, sewage lines etc.	Generally crude and inelegant. Best for continuous features like elevation, soil type, land use, air temperature etc.
Update	Updating database is time-consuming.	Updating database is easier.

Table 3.2 Relative advantages and disadvantages of raster and vector data model

Data structure	Advantages	Disadvantages
Vector	1. Vector data model maintains the accurate geographic location of data	1. The location of each vertex has to be stored explicitly.
	2. The original resolution and form is retained without any generalization.	2. A complete topology has to be built up which is often process-intensive and requires extensive data clearing. Any updating or editing of the vector data requires a rebuild of topology.
	3. The graphic output is aesthetically sound and pleasing.	3. Alogrithms of manipulative and analysis functions are complex and may be process intensive. This inherently limits the functionality for large data sets, e.g. a large number of features.
	4. Topology information helps to make complex operations like proximity, network analysis etc., easier.	4. Continuous data, such as elevation data, temperature profile, can not be effectively represented in vector form. When the data is sparse and scattered, intopolation leads to much data generalization and errors.
Raster	1. The geographic location of each cell is implied by its position in the cell matrix. Accordingly, other than an original point, e.g., bottom left corner, no geographic coordinates are stored.	1. The cell size determines the resolution at which the data is represented.
	2. Due to the nature of the data storage technique, data analysis is usually easy to program and quick to perform.	2. It is difficult to represent linear features adequately depending on the cell (Fig. 3.14 a). Thus proper resolution and accurate network linkages are difficult to establish.
	3. The inherent nature of raster maps (one attribute maps) is ideally suited for mathmatical modeling and quantitative analysis.	3. Processing of associated attribute data may be cumbersome if large amounts of data exist. Raster maps inherently reflect only one attribute or characteristic for an area.
	4. Discrete data, e.g., bore-well location can be accommodated equally well as continuous data, e.g., ground water level and facilitates the integrating of the two.	4. Since most input data is in vector form, data must undergo vector-to-raster conversion. Besides increased processing requirements this may induce data integrity concerns due to generalization and choice of inappropriate cell size.
	5. Grid-cell systems are compatible with raster-based output devices, e.g., electrostatic plotters, graphic terminals.	5. Most output maps from grid-cell systems do not conform to high-quality cartographic needs.

3.7.1 Vector to raster conversion (Rasterization)

Rasterization refers to conversion from vector to raster data. As explained above, Raster format is more convenient to produce color coded polygon maps such as land use or cropping system maps, while map digitizing in vector format is more easier to trace only the boundary. Rasterization is also useful to integrate other GIS data with remote sensing data because remote sensing images are in raster format. This process finds a set of pixels that can approximate points, lines and polygons of the vector coverage. In the original vector coverage locations of objects are precisely defined within the coordinate system. The raster version of vector coverage can only approximate the original locations, and the level of this approximation depends on pixel resolution. A typical rastering algorithm first finds a set of pixels that approximate the locations of nodes in the vector coverage. Then lines joining these nodes are approximated by adding new pixels from one node to the next one in a stepwise fashion.

The rastering process of a linear object produces initially a jagged imprints, differing in its thickness line. These effects are referred to as 'aliasing'. These undesired outcomes can be subsequently corrected by procedures called anti-aliasing. When rastering an arc, the main goal is to approximate its shape with pixels as close as possible retaining their continuity. This process is normally conducted in one direction between the pixels representing start and end of the line. Rastering algorithms estimate the location of the first pixel and then, based on the direction of the line, add pixels one by one to approximate the rest of the line. If the full angular information about the line's direction is applied, rastering calculations become quite complex and time consuming. In order to speed up this process ranges of possible line directions can be represented by integer values, which allow for much faster calculations, hence more efficient rastering process.

The procedures used in rastering of polygons are referred to as polygon scan conversion algorithm. The process begins with the establishment of pixel representations of points and lines that build the polygon. The pixel inclusion criteria is set up to check which of the pixels close to the polygon's edge should be included and which one should be rejected. Then the pixels inside of the polygon are assigned the polygon's identifier. Pixels are usually assigned the 'vector' value based on two different techniques: (i) Central point rastering—the centre of the pixel lies within the polygon and (ii) Dominant unit rastering—pixel gets the value of a polygon that 'covers' the largest proportion of that pixel. In case of a simple conversion all newly created pixels get the same value that comes from the original polygon. In more complex transformation processes, where modelling of original data is involved, values of pixels representing one polygon may vary. The process of converting vector objects to raster pixels uses a range of algorithms that are beyond the scope of this book, however the process described above constitute the basic methodology used in such conversions.

Here is an example of how a simple algorithm for calculation of trapezoid area can be applied to convert vectorized polygon to rasterized polygon with grid cells as shown in Fig. 3.15 (Murai, 1998). If vertical lines are dropped to the x-axis from two adjacent vertices, a trapezoid will be formed.

The area of trapezoid is given by

$$Ai = (x_{i+1} - x_i)(y_i + y_{i+1}) / 2$$

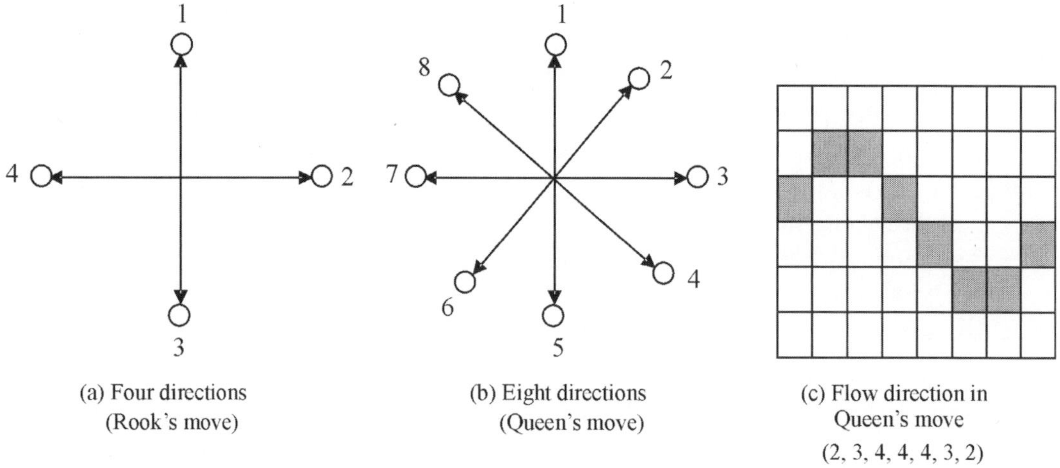

Figure 3.15: An example of rasterization

The sum of all trapezoids will give the area of the original polygon. Using this algorithm, the grid cells in the polygon are easily identified as shown in Figure 3.16.

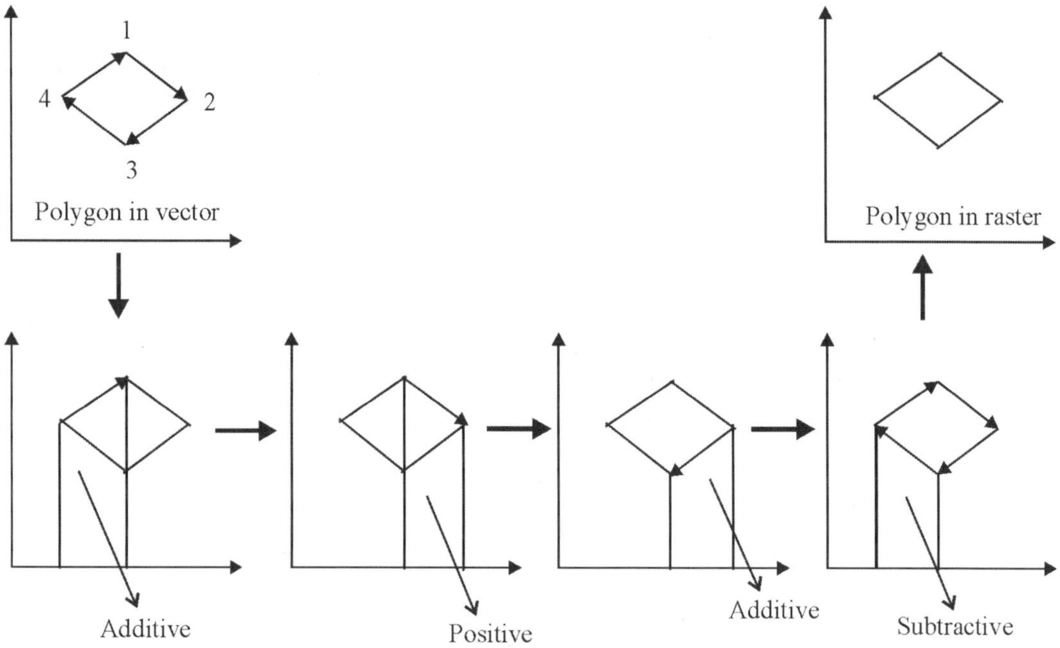

Figure 3.16: Calculating area of a traperoid

3.7.2 Raster to Vector Conversion (Vectorization)

Vectorization refers to the conversion from raster to vector data, which is often called raster vector conversion. In many cases raster images available from such sources as satellite imagery or scanned-in maps must be converted to the vector format for further analysis and/or mapping. (In fact, scanning is a very common procedure used for transforming older hard copy maps into a digital format). Vectorization is not very easy as compared with rasterization, because vector format needs topological structure, for example, direction of line or chain, boundaries and nodes of polygons, order of series of chains that form a polygon, left and right polygon ID of a chain and so on. Vectorisation process constitute the following stages:

1. Creation of Bilevel Images Image is modified so it shows the main feature (foreground) in one shade and all the rest of the features (background) in another shade, e.g., black and white. Various shades of grey are initially present on the raster image. All these shades represent different pixel values.

2. Edge Enhancement This procedure can be applied to introduce more contrast to such image, or to put it in other words—to change the values of pixels so that two distinct groups of different values are present. When edge enhancement step is completed a threshold value is established by viewing the histogram of pixels' values. Pixels on both sides of this threshold value are assigned with two different values, e.g., 1 for the foreground, 0 for the background.

3. Filling This is a process of "closing" line features after the step of bilevel re-classification. In other words, if a line of pixels is incorrectly divided by the background value pixel(s), these pixels are assigned the line pixels' value.

4. Skeletonisation of Line Features (Thinning) If a desired vector representation of linear raster features is a line (arc), some pre-conversion processing of this raster image is usually required. In this process the edge pixels are removed in order to represent a line feature by an array of single yet connected pixels.

5. Vectorizing At this point vector lines can be created and defined geographically at the locations of nodes. Many superfluous vertices are also usually created at this stage.

6. Collapsing Some raster features with substantial area can become complex vector coverages which are then simplified by reducing the initial number of nodes, lines and polygons.

8. Smoothening of Arcs The previous step creates a rugged line for pixel coordinates are not as precise as the node ones. Algorithms are used at this point to smooth the arcs and get rid of unnecessary vertices.

The following example (Fig. 3.17) shows a simple vectorization process with a simple algorithm in which the original image in raster format is converted to vector data through thinning and chain coding. This algorithm is useful to convert raster image to vector data with the coordinates, but it is not sufficient because the algorithm will not build topological structure. Raster vector conversion with automatic building of topology is possible if 2×2 window is continuously moved along the boundary from a node. In order to automate raster vector conversion as much as possible, a clean image without noises or unnecessary marks should be scanned in the beginning.

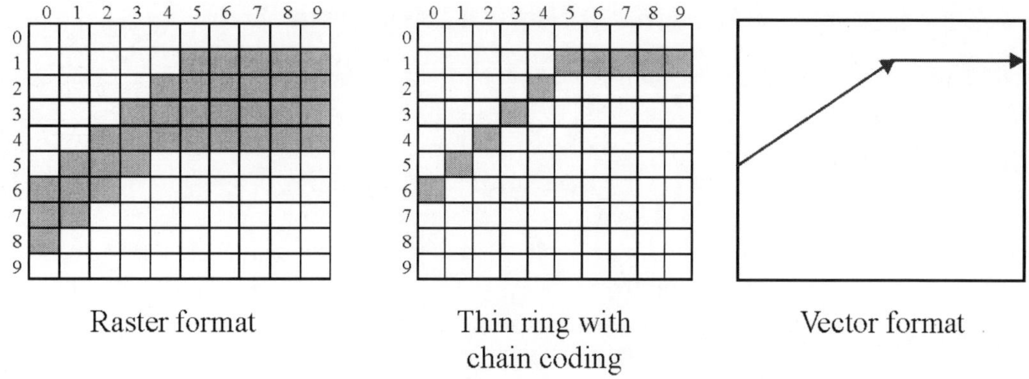

Figure 3.17: An example of vectorization

Chapter 4

Input of Map Data

"The good cartographer is both a scientist and an artist. He must have a thorough knowledge of his subject and model, the Earth.... He must have the ability to generalize intelligently and to make a right selection of the features to show. These are represented by means of lines or colors; and the effective use of lines or colors requires more than knowledge of the subject - it requires artistic judgment."

Erwin Josephus Raisz (1893 - 1968)

4.1 Required Data for a GIS

So far we have developed a conceptual framework of what GIS means and how we can make use of it to describe our complex real world situation. Now we need to examine how practically this can be done with the help of tools, methods or techniques, and how to get our map-information digitally encoded into the system. For all practical purposes, the major data source is usually a map. As explained in the last Chapter, *real* (paper) map needs to be converted to *virtual* (digital) map to work with a GIS. The process of this conversion is called 'geocoding'. Generally, we are well acquainted with maps drawn or printed that are available everywhere—in books and Atlas, as wall-hangers and street guides and so on—something to give us a specific information about a particular area (e.g., population density map of India, cropping pattern map of a farm area) or a drawing of the Earth's surface, showing the shape and position of countries, political borders, natural features like rivers and mountains, and artificial features such as roads and buildings (e.g., Indian political map, map of India with historical places, world map) or a very simple plan which shows a direction of travel between one place and another (road guide maps). Whatsoever the spatial and attribute information they contain, these maps will be of no use to a GIS unless they are converted into computer readable format (digital). Entering spatial information into the system is one of the "core functions" of any GIS. In any GIS project, collecting and getting data into computer is the most expensive and difficult part, sometimes taking up to two-third of the total time. As geospatial data acquisition or input in digital format is the most expensive and the procedures are quite time consuming, the sources of data should be carefully selected for specific purposes. There are data, which are readily available in digital format like remote sensing or GPS data. But most of the data, even in this geoinformation era, are available only in the form of paper maps and in most of the cases one has to convert these analogue maps into digital format to start with any project in a GIS platform. Since GIS has its applications starting from crop production analysis to urban development planning, the source materials also vary widely. Even for a single GIS project, the range of required data can be very vast. Nevertheless, it is essential to find the existing map data first and then to go for creating data. This chapter discusses the available sources of data, methods available to get these data into computers, their editing and storage before using them in future for some applications. It must be remembered that care should be taken at the first to avoid the time taken for editing and corrections later.

First let us have a look at the major data sources which are widely used (Fig 4.1). One must have a choice of data subjected to the relative cost involved and accuracy needed, as listed in Table 4.1.

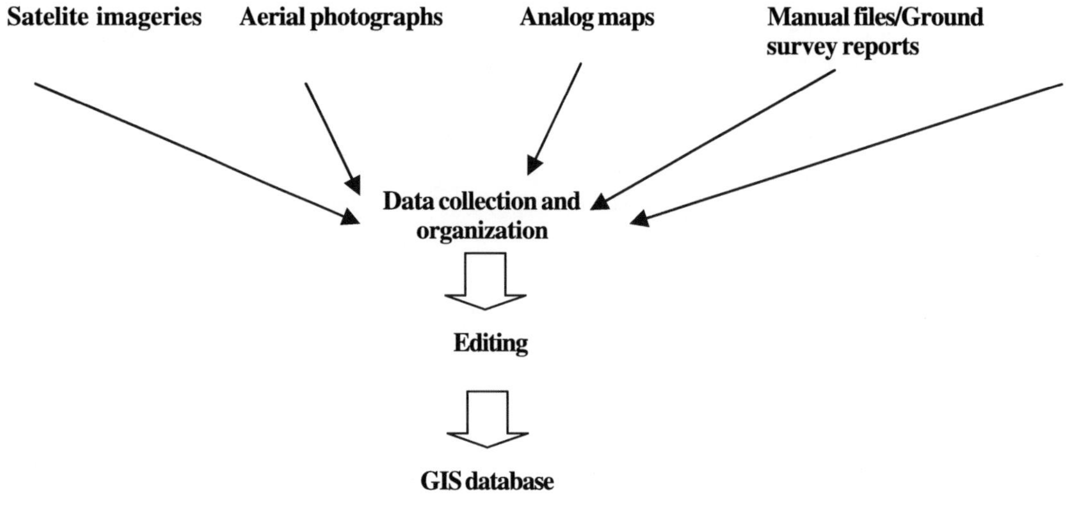

Figure 4.1: Major data sources for GIS

4.1.1 Analog maps

Topographic maps with different terrain features (contours, drains, spot heights) and thematic maps with defined object classes are digitized manually through digitizers or semi-automatically through scanners. Problems of analog maps are lack of availability, out of date, inconsistency in map production, time, inaccuracy etc. These maps are mostly available in public domain. Many

Indian national agencies like Department of Census (demographic and socioeconomic data), Survey of India (SOI) (topographical and land use/ land over data), Geological Survey of India (GSI) (geology and topography), Central Ground Water Board (CGWB) (groundwater prospects, quality of groundwater), National Atlas Thematic Mapping Organization (NATMO) (soils, topography), All India Soil and Land Use Survey (AISLUS) (soils), National Bureau of Soil Survey and Land, Planning (NBSS&LUP) (Soil), National Remote Sensing Agency (NRSA) (satellite data as FCC and digital format) and many state level agencies collect and store data in the form of maps except some restricted areas, the maps of which are not generally available. Nowadays, there are many organizations, which also maintain records (in the form of analog maps) ranging from land resources to pollution levels regionally or across the country. These maps vary widely in scales, but they can be used by any person after getting permission from the competent authority. Countries like US, Canada, and many European countries do maintain maps sometimes more systematically than us, providing availability of maps at ease.

There are some maps drawn on paper form available in books and atlases. These maps are often excellent form of information and can be used after suitable modifications depending upon the objectives of their use.

Table 4.1: Choice for different data sources

Data source	Method	Equipments	Accuracy	Cost
Analog maps	Manual digitizing	Digitizer	±0.1 mm	Cheap
	Semi-automatic scanning	Scanner	±0.1 mm	High
Areal photographs	Analytical photogrammetry	Stereo plotter (analog)	± 10 cm	High
	Digital photogrammetry	Digital photo workstation	± 10 cm	Very high
Satellite imageries	Visual interpretation	Image zoom scope	± 30-50 m	Cheap
	Digital image	Image processing software	± 10-30 m	High
Ground survey	Field survey and analysis	Total station, GPS	± 1 cm	Very high
Reports	Entry through keyboard	Keyboard, mouse	Varies	Cheap

4.1.2 Aerial Photography

According to Curran (1989) and Mather (1991), aerial photography is the process of acquiring images from a position above the Earth's surface without being in contact with the object of interest, providing a 'snapshot' of the Earth's surface at a particular instance of time. Ranging from visible band (black/white and color images) to infrared sensor aerial photography, analytical or digital photography is extremely useful in monitoring changes such as land use/ land cover, vegetation and river dynamics, drainage network, urbanization, etc., and thus offers a very good source of data for GIS.

The photographs obtained aerially must be georeferenced first for their subsequent use. The photographs may be vertical (when taken directly below the aeroplane or oblique (if taken at an angle). As the oblique photographs cover larger area it is cheaper than vertical ones, but have less usefulness for GIS. Even for the vertical photograph, the scale at which the photograph is taken is the most important point to consider. As the height of the flight increases from the ground, the scale difference between the centre and the edges of the image increases and as a result, the edges of the photograph get more distorted (Fig. 4.2). The data of the photograph should match with other observations (supportive data or secondary data). Again, as the shades of the objects vary depending on the Sun's position, the time of the photograph should also be taken care of.

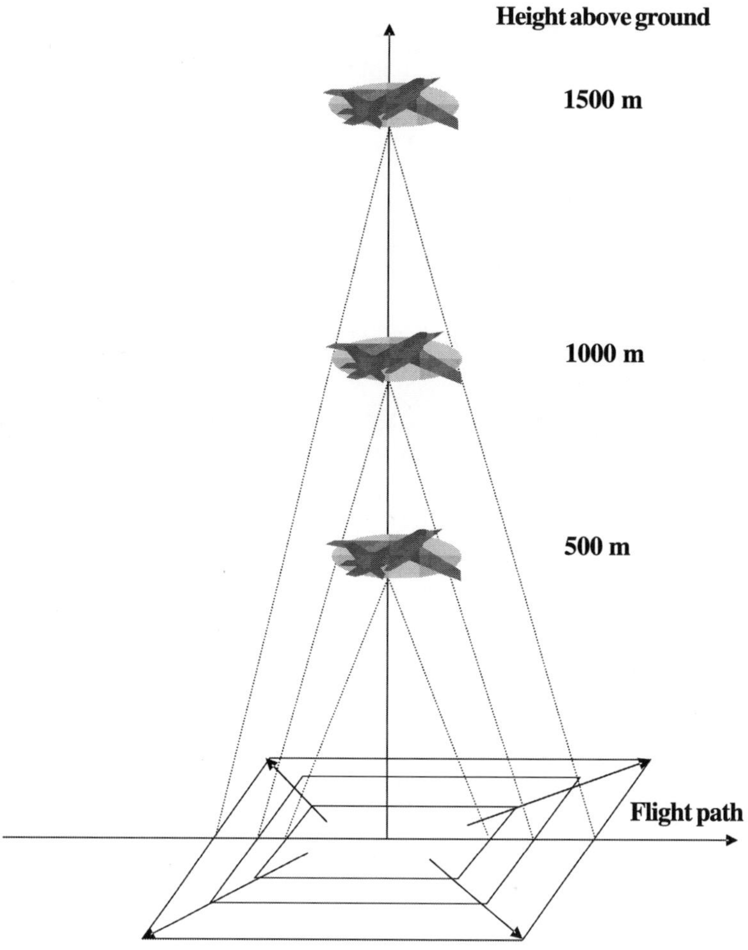

Figure 4.2: Aerial photographs with varying scales (Source: Heywood et al., 2000)

Aerial photography (analytical or digital) data is expensive but is often considered the best method for updating, even in time of availability of satellite data. The characteristics for which aerial photographs may be of immense value are listed here (Curran, 1989; Heywood et al., 2000).

1. Wide availability
2. Relatively inexpensive in comparison to other remotely sensed imageries
3. Detailed data
4. Wide area view
5. Time-freezing ability
6. High spectral and spatial resolutions
7. Three-dimensional perspective
8. Only source of data for objects like location and extent of new housing or forest fire, where spatial information is difficult to be obtained from other secondary sources.

4.1.3 Satellite data

Sensors-on-board-satellite over a wide spectrum of electromagnetic wave band collects information on objects situated at a distant place and relay the information in the form of electric signals to Earth receiving stations. These signals are processed by computers to produce an image, which may be in both digital and analog formats. Detailed account of the process of data acquisition by satellites and their subsequent integration through GIS are presented in Chapter 10. Satellite data in the form of imagery (False Color Composite, FCC) or digital data format are available for land use classification, digital elevation model (DEM), updating highway network, mapping soil moisture, identifying fishing zones and many other applications. But the image map scale would be around 1:50,000 to 1:100,000. Presently, high-resolution satellite imageries are available with ground resolution of 1 to 3 meters that can produce 1:25,000 topomaps. Some of the advantages of satellite data over the other sources of data which makes it a beautiful source of information for GIS applications are synaptical view and repetitive coverage over a large area, near correctness and completeness of data, nondestructiveness in nature and a tool for obtaining information virtually impossible for any other tool or technique.

4.1.4 Ground survey with GPS

In many of the cases data required for the study do not exist and need to be collected *in situ*. The same is true when some additional data are needed either to verify satellite data or other data or to complete the required datasets. In these cases traditional manual survey techniques are followed. Engineering survey with chains, plane tables, levelers and theodolites or soil survey with augers, samplers or groundwater survey using peizometers and geophysical techniques are some of the examples. Nowadays total station i.e., high precision theodolites with electronic distance metering (EDM) with a data logger or modern digital recording machines and a hand-held laser range finder are available to collect data faster and automatically. The most recent advancement in the field of survey is Global Positioning System (GPS), which modernize the ground survey (see section 4.4). It is very accurate but expensive to cover wide areas.

4.1.5 Reports and publications

Socio-economic data are usually listed in the reports of statistics and census with respect to administration units. Many collateral data are also available in the form of tables or graphs. There are data that has already been collected by some sources and can nowadays be even downloaded from the Internet, or obtained on a variety of electronic media, such as CD-ROMs.

4.2 Existing Datasets

Most of the different kinds of maps available in the public domain are printed and have to be translated into digital or number form to use them in a GIS. In Indian context not many data are available in digital format and one has to digitize or scan those maps to get them into a computer. Unlike this in US, map data created by the federal government are available in digital format. The Freedom of Information Act guarantees every American the right to get copies of digital map data used by the federal government, except for certain sensitive areas. Nowadays commercial companies have started selling data products and may also commercially scan and digitize a map for one's use. This type of services though well established in US and other developed countries, are very few in India.

4.2.1 Data on the networks

In this era of internet, what may be the best possible way to find any data available worldwide, other than the World Wide Web (WWW)? Agencies that supply data have a gateway or server from where the data can be downloaded. Some prominent WWW gateways for data are United States Geological Survey's (http://www.usgs.gov/), National oceanic and Atmospheric Administration's (http://www.noaa.gov/) and US Bureau of Census (http://www.noaa.gov/). Though India is still at its infancy in putting its own data on networks, there are attempts to place data products of India on networks, which can mostly be freely downloaded. Some of them are as follows.

4.2.1.1 Poverty and Human Resources Division Policy

Poverty and Human Resources Division Policy, Research Department, The World Bank (http:/ /www.worldbank.org/poverty/data/indiadata.htm) This site contains database comprising 30 spreadsheets and 89 text files (ASCII) that are grouped into six subject blocks with detailed statistics on a wide range of topics in India. The data are presented at the state level and at the all India level separately. The database uses published information to construct comprehensive series with period coverage from 1950 to 1994. The list of variables include expenditures (distribution), national accounts, price wages, population and rainfall.

4.2.1.2 Statistical data of India (www.indiastat.com)

It provides a huge amount of India-specific socio-economic data and figures gathered from various secondary level authentic sources. In this server over half-a-million pages of statistical data have been qualitatively analyzed, condensed and presented in a user-friendly format. This exhaustive and methodically compiled data can easily be accessed. The data can be downloaded in MS-Excel/HTML formats. There are other associated sites like indiaagristat.com, indiaeconomystat.com, indiaindustrystat.com, indiatourismstat.com, etc. (sector-specific sites), and delhistat.com, biharstat.com, keralastat.com (state-specific sites).

4.2.1.3 The Census Data Online (http://www.censusindia.net/)

This site belongs to the Registrar General and Census Commissioner, India and provides important census data on the people of India (demographic and socio-economic aspects) at the country level at one place. User can search for data on a number of topics collected from the official Indian census. Maps and other data pertaining to India are also included in this site.

4.2.1.4 Infotech GIS Data for India

Infotech GIS, an ISO 9002 accredited GIS conversion services company offers services with a capacity to do 20,000 hours of work per month in Arc/ Info, Intergraph, MapInfo, and Atlas GIS formats. This company develops software for GIS users in Windows environment using state-of- the-art programing technologies.

4.3 Input of Geospatial Data

Input devices are of many types, most of which are used for GIS data input. But to call an input system effective, it must fulfil four major criteria. According to DeMers (2000), these are: (a) The system must be designed to transfer both entities and attribute data into the computer; (b) It must have one of the two ways to view the real worldeither as raster (grid cells) or vector entities; (c) it must have a form of cartesian grid; and (d) The system must ensure that the input data is saved and can be retrieved for editing only to save it again.

The modes of data input may broadly be categorized into three groups—manual devices (e.g., digitizers, computer mouse), automated devices (e.g., scanner) and conversion from other sources. These modes along with the methods/ instruments commonly followed for input of geospatial data into computer system are described below.

4.3.1 Manual devices

In this case the user directly manipulates a device whose location is recognized by the computer. In the earliest GIS packages, maps were encoded and entered by hand. This involved producing a square grid on a clear plastic material (Mylar) and then transforming (Typing) the data grid by grid, into other computer. This process of manual geocoding involved lot of time, labor and errors. As the computer systems developed, manual map encoding process got replaced by automation, popularly known as 'digitization'. Digitization process is similar to drawing maps manually, here the *pencil* is replaced by the *cursor,* which is being traced over lines to feed data into the computer. Though this geocoding process involves mechanical device as *cursor*, it is called *semiautomated digitizing* as human operator is also involved in the process ('scanning' is another operation which is fully automatic).

4.3.1.1 Digitizer tablet

Digitizers are the most common devices for extracting spatial information from maps and photographs. Tablet digitizer with a free cursor (often called *puck*) connected to a personal computer is used for digitizing spatial features with the planimetric coordinates from analog maps (Fig. 4.3). The analog map is placed on the surface of the digitizing tablet. The position of an indicator (it may be a pen-like stylus or a cursor, a small flat plate with a cross-hair) as it is moved over the surface of the digitizing tablet is detected by the computer and interpreted as pairs of (x, y) coordinates. Frequently, there are control buttons on the cursor (more the sophistication), which permits control of the system without having to turn attention from the digitizing tablet to a computer terminal. The size of digitizer usually ranges from small (paper-size) (A3) to large (A0) size. Early digitizers (ca. 1965) were backlit glass tables, where a magnetic field generated by the cursor was tracked mechanically by an arm located behind the table. The arm's motion was encoded, coordinates computed and sent to a host processor. The first solid-state systems used a spark generated by the cursor and detected by linear microphones. Contemporary tablets (accuracy ~ 0.1 mm) use a grid of wires embedded in the tablet to generate a magnetic field, which is detected by the cursor. A good digitizer tablet must be stable and produce precise results repeatedly, limit should be as minimum as possible. Here are some points on the digitizing process.

Figure 4.3: Operation of digitizing tablet

A. Steps in Digitizer Operation

The steps are as follows:

1. First to ensure that a proper connection is made between the digitizer tablet and the computer workstation or a PC terminal.
2. The map has to be firmly fixed or taped to a digitizing tablet. There should not be any folds or wrinkles during the fixing and the surface of the map should be flat and smooth to enable free movement of the cursor or stylus over it.
3. A coordinate system for the map has to be chosen (mostly UTM or latitude/longitude). Sometimes the map units (inches or millimeters) are used for precise matching between the digitized map and its source map.
4. Control points or *tics* are placed at four corners of and within this map sheet in such a way that they produce a perfect square/rectangle. These points should be easily identifiable like intersections of major streets, railway lines and/or roads, spot heights, points on coastline etc. The coordinates of these points will be known in the coordinate system to be used in the final database, e.g., latitude/longitude, x-y, State Plane coordinates etc. The control points are used by the GIS system to calculate the necessary mathematical transformations to convert all coordinates to the final system. Thus, it is imperative that the control points should be very precise to achieve the best of the accuracy. In order to achieve more accuracy the control points may be digitized repetitively and the average values should be taken.
5. To start with digitization, the control points should carefully be selected and their coordinates are entered interactively. During the process the map should not be removed and fixed again, and the whole digitization process should be completed in one go.
6. Map contents (features) are traced out using the cursor in either *point mode* or *stream mode* for a short time interval, a process called 'digitization'. The point mode is being mostly used where the operator identifies the points to be captured explicitly by pressing a button. In stream mode the points are captured at set time intervals (typically 10 per second) or on movement of the cursor by a fixed amount. Here the operator generates large number of points, many of which may be redundant. There is another mode called *point select mode* which provides an option of switching between point and stream modes.
7. Points are entered one at a time and then its attributes are entered. In other way the attributes for all the digitized points can be entered later on. Similarly lines are entered as strings of points and should be terminated either by hitting some button or by digitizing the point in a preset menu area to determine the end-node. Areas are digitized as lines with an automatic closure (*snapping*).
8. Editing is performed to check and remove errors such as small gaps at line junctions, overshoots, duplicates etc., (Fig. 4.4) This is to make for a clean dataset without errors.

B. Limitations in Digitizing Using the Tablet

As the digitization process involves human component, errors are likely to creep in, which may vary depending on the individual. It also depends on the map quality as well as selection of the control points. Over all there are some common problems encountered while digitizing. According to Marble et al. (1984), these are:

(i) Paper maps are unstable. It may stretch or shrink as times passes, which makes the newly digitized points slightly off in their location from the previous points;

(ii) The map itself may contain errors as these are mostly meant to display information, and may not necessarily be accurate record locational information.

(iii) Discrepancies across neighboring map sheets will produce disconnectivity and errors in total GIS database; and

(iv) Operators are likely to make errors and mistakes while digitizing (Fig. 4.4).

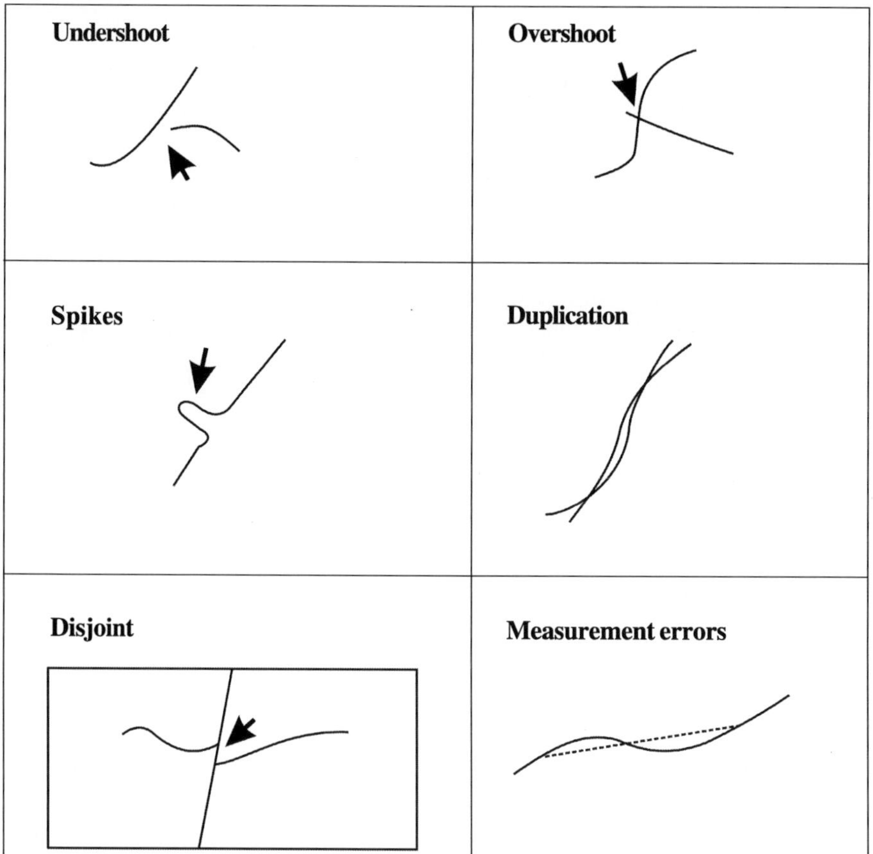

Figure 4.4: Some typical errors in digitization

C. Some Guidelines to Reduce the Errors

1. One must ensure that the printed map is in good quality. The map should be in good physical condition and does not have any folds, wearing and tearing or wrinkles.
2. The map should be firmly fixed on the table and once removed, the control points must be re-entered when the map is affixed again. Its surface should always be flat and smooth.
3. Make out a plan before digitization. Decide what features are to be digitized and in what sequence. Complete one feature, and then start another.
4. Locate *tic* points correctly (the *tic* point should be correctly identified under real world situation). Assign them unique numbers. For multiple maps, ensure that the *tic* pints are matched properly.
5. Decide the standard names for different features before digitization begins.
6. Digitize contours, roads, rivers etc., to some extent beyond the boundary to ensure that the whole area has been represented. This will help for some spatial analysis, and to have a check on the boundary.
7. Ensure that there is no missing data (i.e., all the required features have been fully digitized).
8. Check whether all the polygons close and each polygon has only one labeling point with a unique identifier.
9. Establish the topology correctly.

4.3.2 Input through Keyboard

Keyboard entry is basically for non-spatial attributes and occasionally locational data. Generally this is done along with the map digitization process to generate a complete GIS database (both spatial and attribute data). This method involves a lot of error and its utility is also very limited. Sometimes, for some limited number of features for which coordinates are known, the keyboard can be used for the entry.

4.3.3 Scanner

Scanner is a fully-automated input system with little intervention from the user. Nowadays scanner is a common item in most computer facilities and the users can quickly input graphic and text information directly into a computer in *raster* format. If a vector product is required, line-tracing algorithms must be used to convert the raster to vector. The quality of this transformation is related to the quality of the scanner and the quality of the base map being scanned. The scanned data (digital image data) are usually integer-based with one byte gray scale (256 gray tones from 0 to 255) for black and white image and a set of three gray scales of red (R), green (G) and blue (B) for color image. Following four types of scanners are commonly used in GIS (Fig. 4.5).

Electromechanical Scanner It is also called drum scanner since a map or an image placed on a drum is digitized mechanically with rotation of the drum and shift of the sensor as shown in Fig. 4.4. A scanner head containing a light source and photo detector reads the reflectivity of the target graphic, and digitizing this signal, creates a single column of pixels from the graphic. The rotation of the drum is same as the sensor's movement along the frame. The scanner head moves along the axis of the drum to create the next column of pixels, and so on through the entire scan. This type of scanning is expensive and slow, but very accurate.

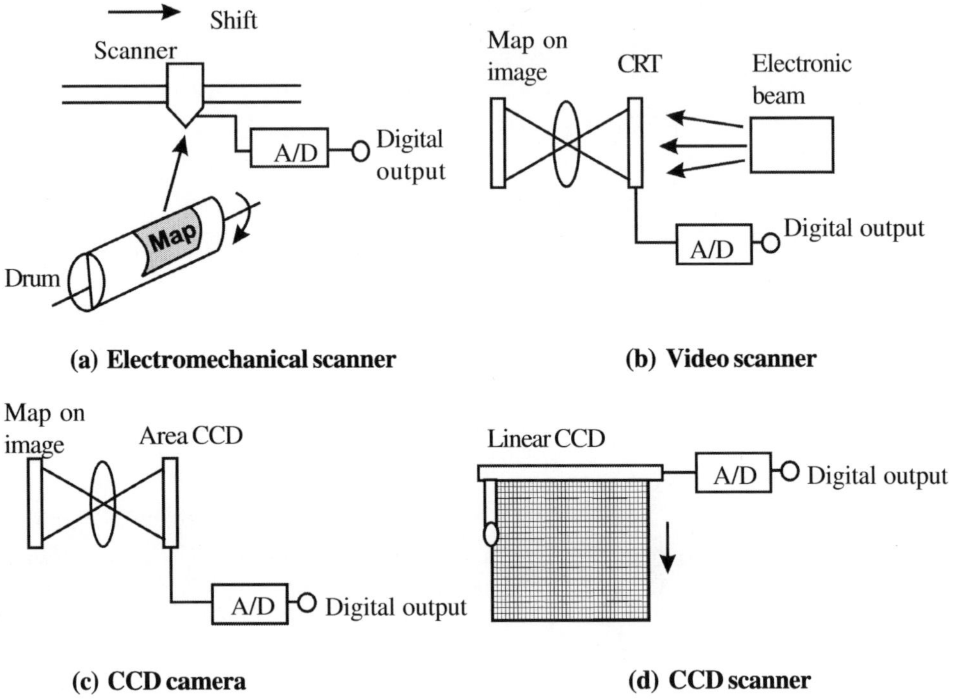

Figure 4.5: Major types of scanner

Video Scanner Video camera with CRT (cathode ray tube) is often used to digitize a small part of a map. These are essentially television cameras, with appropriate interface electronics to create a computer-readable dataset. This is extremely fast and relatively cheap but not very accurate. This type of scanner produces a raster array of brightness (or color) values, which are then processed much like any other raster array. A typical data arrays from video scanners are of the order of 250 to 1000 pixels on a side. But this data input procedure has typically poor geometrical and radiometrical characteristics, including various kinds of spatial distortions and uneven sensitivity to brightness across the scanned field. Thus, video scanners are difficult to use for map input because of the problems associated with distortion and interpretation of features.

Charged Couple Device (CCD) Camera Area CCD camera (called digital still camera) instead of video camera is more convenient to acquire digital image data. It is more stable and accurate than video camera.

CCD Scanner Flat bed type or roll feed type scanner with linear CCD is now commonly used to digitize analog maps in raster format, either in monotone or in color mode. It is very much accurate and the cost involved with it is moderate to high. These scanners may be of small to large size depending on the size of the document to be scanned. The larger format commercial scanners produce a very high resolution involving lot of memory/stage space. A comparison between different scanners is presented in Table 4.2.

Table 4.2: Comparison of different scanners.

Type	Resolution	Accuracy	Cost
Mechanical scanner	Very high (25–100m)	Very high	High
Video camera	Low (500 × 500 pixels)	Low	Cheap
CCD camera	Medium (500 × 500–4,000 × 4,000)	Medium	Medium
CCD scanner	Very high (300-600 dpi)	Very high	Medium to very high

Though scanners are more useful than digitization, large-format commercial scanners cost very high and out-of-reach to many GIS users. For large complex maps, manual digitizing may be the only option, even in the near future. There may be distortions along the edge of the scan area, as most of the desktop scanners are not designed for accurate planimetric applications. Most maps include many graphic and text features, annotations along with the input features like contours, roads and drains. There is no possible way to scan the required input features only (color, roads) without scanning other associated features at the same time, thus the post-scan data clean up process becomes very tedious and time consuming. Many times the lines are not clear and wide enough to be detected clearly by the scanner. Maps should be clear of text and other graphical features that are not required. Vectorization process also brings in many errors depending on the quality of the line-trace algorithm.

4.3.4 Aerial photogrammetry

Photogrammetry is an indirect method of measurement and derives geometrical parameters of *remote objects* from the photographs taken from the aeroplane. Depending on the object-class two cases can be distinguished—arial photogrammetry and terrestrial or close-range photogrammetry if the object is the earth surface with topographic features and in the latter any other object, e.g., a building, a human face, industrial artifacts, etc. The imaging process is mathematically formulated by a perspective transformation, which gives the relation between the position of a point in the photograph (described by "image coordinates") and its actual object coordinates (x, y, z). The goal of photogrammetric plotting is the inversion of the perspective, i.e., the computation of object coordinates from measured image coordinates.

(a) Analog photogrammetric instrument

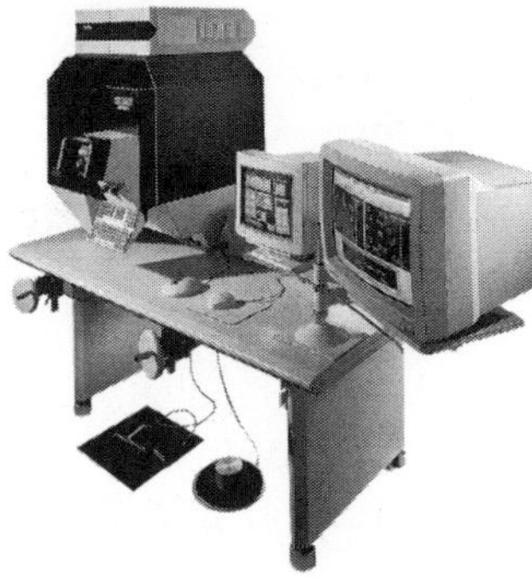

(b) Analytical stereo plotter

(c) Digital photogrammetric workstation

Figure 4.6: Instruments used in photogrammetry

Through the photogrammetry the output may be either two-dimentional or three-dimentional vector data, or raster data (orthophotograph). The methods of measurements of vector data are analog stereo plotting (analog pictures, analog instruments), analytical stereo plotting (analog pictures, analytical stereo plotter) and digital plotting (digital pictures, digital mono plotting, digital stereo plotting). Some of these instruments are shown in Fig. 4.6.

Though aerial photogrammetry is rather expensive and slow in air flight as well as subsequent photogrammetric plotting and editing, it is still very important to input accurate and up-to-date spatial information. Aerial photogrammetry needs a series of the procedures including aerial photography, stereo plotting, editing and output as shown in Fig. 4.7. There are two types of aerial photogrammetry.

Analytical Photogrammetry Though computer systems are used for aerial triangulation, measuring map data, editing and output with pen plotter, a stereo pair of analog films are set up in a stereo plotter and the operator manually reads terrain features through stereo photogrammetric plotter called analytical plotter.

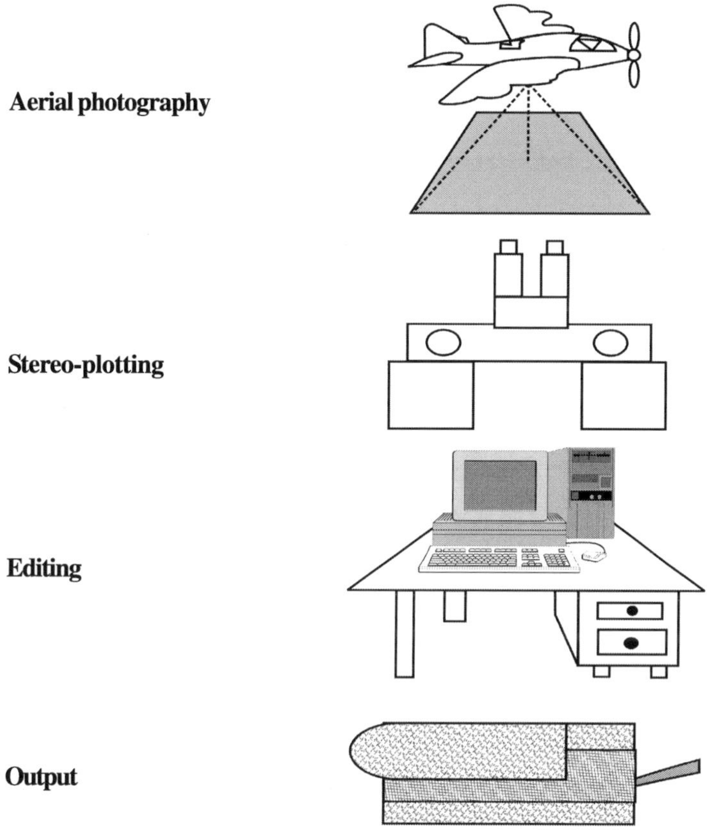

Figure 4.7: Aerial photogrammetry

Digital Photogrammetry In digital photogrammetry, aerial films are converted into digital image data with high resolution (5-25 mm). Digital elevation model (DEM) is automatically generated with stereo matching using digital photogrammetric workstation. Digital ortho-photo and three-dimentional bird's eye view using DEM will also be automatically created as by-products. Future research must aim for identifying the patterns of houses, roads, structures and other terrain features automatically. Though this method is very expensive there are several advantages connected with using digital images:

(i) Digital soft copy workstations running on standard computers can be used rather than expensive plotting devices requiring special hardware;

(ii) Data transfer is simpler with regard to plotting results because they can be post-processed on the same computer;

(iii) Digital image processing techniques can be used for image enhancement; and

(iv) The same renders possibility of the automation of photogrammetric measurement tasks.

The aerial photographs may be of black and white color, or infrared film. Special aeroplanes are used as the platforms of the sensors of photogrammetry, but sometimes the helicopters can be used. Overlapping photographs are necessary for production of three-dimentional vector data. In this case aerial photographs cover the whole area usually with a 60% overlap along the flight lines and a 20% overlap between flight lines.

However, there are problems connected with image acquisition. For many applications the format of CCD-sensors, which are commercially available, is too small or the sensors are too expensive. In this case or if high precision is required and metric cameras have to be used, one will first get analog photographs. Those images have to be scanned off-line using photogrammetric scanners, which turn out to be a very time-consuming process. In order to achieve precisions comparable to those achieved in analog photogrammetry, point localization has to be done with accuracy better than 1 pixel (sub pixel accuracy).

4.3.5 Satellite imagery

Satellite remote sensing is a modern tool to obtain digital image data of the terrain features in the electromagnetic regions of visible, infrared and microwave. Multispectral bands including visible, near-infrared and/or thermal infrared are most commonly used for production of land use map, soil map, geological map, crop acreage map, forest map etc., at the scale of 1:50,000 to 1:250,000. A lot of earth observation satellites e.g., LANDSAT, SPOT, ERS-1, JERS-1, IRS, Radarsat etc., are available to supply data to be used for various applications (Table 4.3). Synthetic aperture

Figure 4.8: A satellite imagery showing Indian subcontinent

radar (SAR) is now becoming a new technology in remote sensing because SAR can penetrate through clouds, which enables cloud free imagery in all weather conditions. Satellite imagery (Fig. 4.8) is often the preferred choice of imagery for larger regions. Imagery is a powerful visual aid and serves as a source of derivative information such as planimetric and classification schemes to derive information like land use or vegetation. Satellite image has different ground resolutions depending on the sensors used. Since cold war, very high resolution satellite imageries with ground resolution of 1 to 3 meters are becoming available. Such high resolution satellite images are being extensively used to identify individual houses in urban area. The high resolution satellite images has high potential to be applied to urban GIS. More and more choices of satellite imagery are becoming available today and the cost associated with its purchase is dropping. A detailed discussion on remote sensing is given in Chapter 10.

Table 4.3 Some current earth observation satellites

Name	System	Spectral bands	Ground resolution (m)	Main use
Landsat™ (USA)	TM	6 (VIS & NIR)	30	Land use/ land cover
		1 (TIR)	120	
SPOT (France)	PAN	1	10	Topography
	XS	3	20	Land use/land cover
ERS-1 (EC)		SAR (C)	30	Wind, wave and ice mapping
JERS-1 (Japan)	OPS	8	18´24	Land cover, geology and terrain features
	SAR	SAR (L)	18´18	
IRS-ID (India)	MSS	4	23.5	Land use/land cover, vegetation
	PAN	1	5.8	Topography
RADARSAT (Canada)	SAR (C)	SAR (C)	10-50	Ice and snow mapping

4.3.6 Advanced Technologies for Primary Data Acquisition

Several advanced technologies have now become available for acquisition of primary geospatial data as well as digital elevation model (DEM) which are listed here (Fig. 4.9).

Electronic Plane Surveying System An integrated system of total station with automated tracking function, kinematic GPS and a pen computer can replace the conventional plane surveying. Direct data acquisition in digital form, at the field scale will be very useful for large scale applications of GIS data sets, e.g., in the application to cadastre, utility facilities, urban structures etc.

Mobile Mapping System Different sensors such as GPS, INS (Inertia Navigation System), more than two digital cameras, voice recorder etc., are fixed on a vehicle in order to map objects in a close range (for example, center line of highways, utility lines, railways etc.) as well as to determine trajectory of the moving vehicle.

Laser Scanner Airborne laser scanner together with GPS (see Section 4.4) and INS will measure directly the terrain relief or DEM with the height accuracy of 10 cm up to the altitude of 1,000 m.

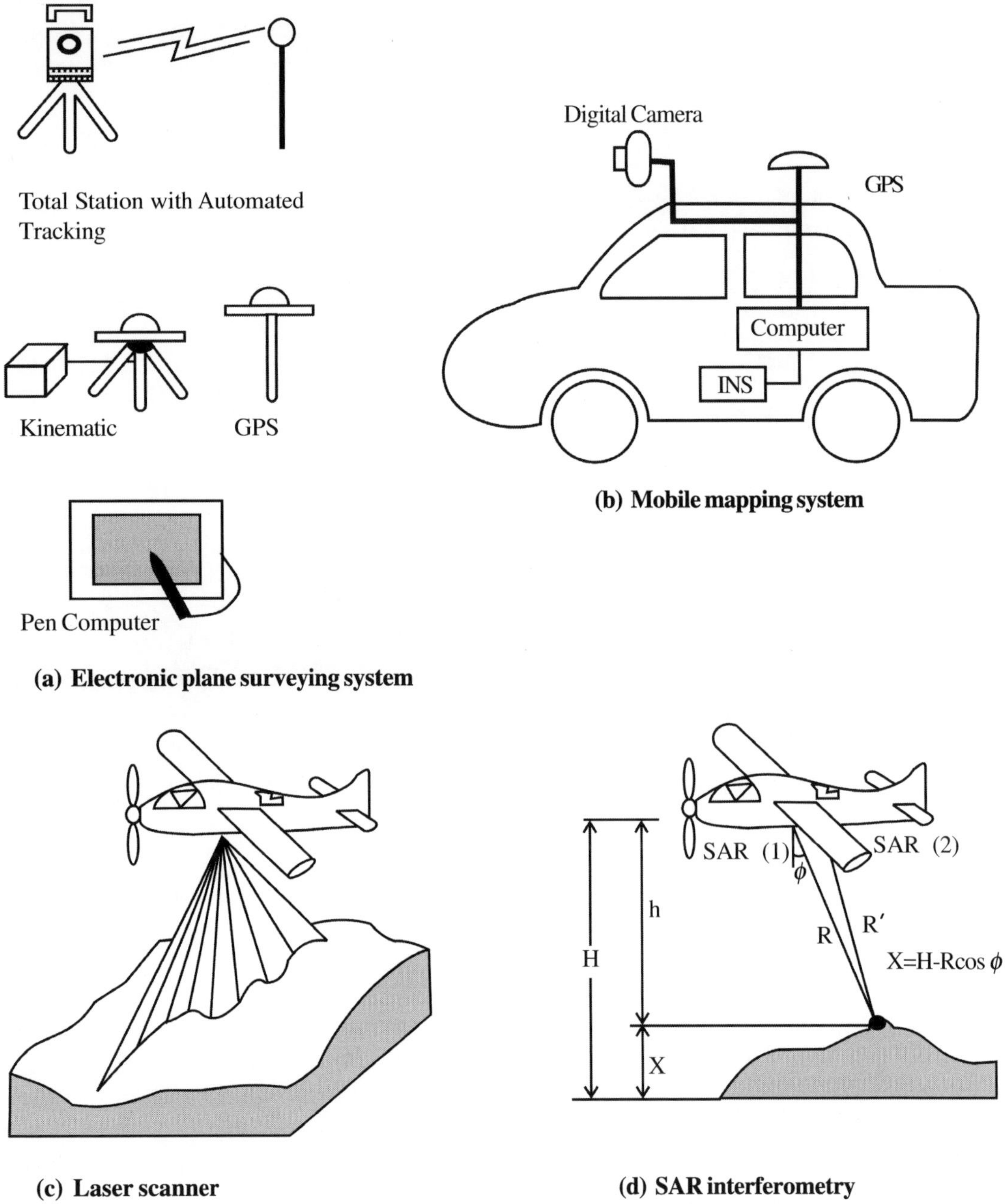

Figure 4.9: Advanced technologies for primary data acquisition

Synthetic Aperture Radar (SAR) SAR interferometry is the latest technology to produce DEM automatically by special interferometric processing of a pair of SAR images. Though still at its infancy, airborne and space borne SAR interferometer are now available if the interferometric condition meets the standard.

4.4 GPS SYSTEM

The Global Positioning System (GPS) is a satellite-based navigation system that can be used to locate positions anywhere on the Earth. GPS provides continuous (24 hours/day), real-time, three-dimensional positioning, navigation and timing worldwide in any weather condition. GPS was originally intended for military applications, but in the 1980s, the government made the system available for civilian use. There are no subscription fees or setup charges to use GPS. Any person with a GPS receiver can access the system, and it can be used for any application that requires location coordinates.

4.4.1 COMPONENTS OF GPS

The GPS system consists of three segments: (1) The space segment—the GPS satellites themselves, (2) The control system, operated by the U.S. military, and (3) The user segment, which includes both military and civilian users and their GPS equipment.

4.4.1.1 Space segment

The space segment is composed of the GPS satellites that transmit signals from space, based on which time and position of the user is measured. The whole set of satellites is called 'constellation'. There are two GPS constellation such as NAVSTAR and GLONASS.

NAVSTAR (**N**avigation **S**atellite **T**iming and **R**anging) constellation operated by U.S. Department of Defense (DoD) is composed of 24 satellites in six orbital planes. The first GPS satellite was launched in 1978. A full constellation of 24 satellites was achieved in 1994. Four satellites move in circular 20,200 km orbit at an inclination angle of 55 degrees to the equator and with a 12 hour period. The high altitude ensures that the satellite orbits are stable, precise and predictable, and that the satellites' motion through space is not affected by atmospheric drag. The GPS satellites are powered primarily by sun-seeking solar panels, with Ni-Cd batteries providing secondary power. On board, each GPS satellite has four atomic clocks, only one of which is in used at a time. These highly accurate atomic clocks enable GPS to provide the most accurate timing system.

GPS satellites transmit two low power radio signals, designated L1 (1575.42 MHz) and L2 (1227.6 MHz). Each transmits on exactly the same frequency; however each satellite signal is Doppler shifted by the time it reaches the user. L1 carries a precise (P) code and a coarse/acquisition (C/A) code. L2 carries only the P code. Civilian GPS uses the L1 frequency. The P code is normally encrypted so that only the C/A code is available to civilian users. The P code when encrypted in known as Y code. The signals travel by line of sight, meaning they will pass through clouds, glass and plastic but will not go through most solid objects such as buildings and mountains.

A GPS signal contains three different bits of information—a pseudorandom code, ephemeris data and almanac data. The pseudorandom code is simply an I.D. code that identifies which satellite is transmitting information. One can view this number on GPS unit's satellite page, as it identifies which satellite it is receiving.

Ephemeris data, which is constantly transmitted by each satellite, contains important information about the status of the satellite (healthy or unhealthy), current date and time. This part of the signal is essential for determining a position.

The almanac data tells the GPS receiver where each GPS satellite should be at any time throughout the day. Each satellite transmits almanac data showing the orbital information for that satellite and for every other satellite in the system.

GLONASS The Russian government has developed a system, similar to GPS, called GLONASS. The first GLONASS satellite was launched in October 1982. The full constellation consists of 24 satellites in 3 orbit planes at a height of 19,100 km, which have a 64.8 degree inclination to the Earth's equator. The orbiting period is 11 hours 15 minutes. The GLONASS system now consists of 12 healthy satellites. Each satellite transmits on two L frequency groups (L1 group is centered on 1609 MHz and L2 on 1251 MHz). GLONASS uses the same code for each satellite and many frequencies, whereas normal GPS uses two frequencies and a different code for each satellite. The GLONASS signals carries both P and C/A codes. Some GPS receiver manufacturers have incorporated the capability to receive both GPS and GLONASS signals. This increases the availability of satellites and the integrity of combined system.

GALILEO Galileo is Europe's contribution to the next generation Global Navigation Satellite System (GNSS). Unlike GPS, which is funded by the public sector and operated by the U.S. Air Force, Galileo will be a civil controlled system that draws on both public and private sectors for funding. The service will be free at the point of use, but a range of chargeable services with additional features will also be offered. These additional features would include improved reception, accuracy and availability. Design of the Galileo system is being finalized and the delivery of initial services is targeted for 2008.

4.4.1.2 Control segment

The control segment consists of five Monitor Stations (Hawaii, Kwajalein, Ascension Island, Diego Garcia, Colorado Springs), three Ground Antennas (Ascension Island, Diego Garcia, Kwajalein), and a Master Control Station (MCS) located at Schriever AFB in Colorado. The monitor stations passively track all satellites in view, accumulating ranging data. This information is processed at the MCS to determine satellite orbits and to update each satellite's navigation message. Updated information is transmitted to each satellite via the Ground Antennas.

4.4.1.3 User segment

The GPS User Segment consists of the GPS receivers and the user community. GPS receivers convert satellite signals into position, velocity, and time estimates. Four satellites are required to compute the four dimensions of x, y, z (position) and time. GPS receivers are used for navigation, positioning, time dissemination, and other research. There are civilian applications for GPS in almost every field, from surveying to transportation to natural resource management to agriculture. Most civilian uses of GPS, however, fall into one of four categories: navigation, surveying, mapping and timing.

4.4.2 Working principles

Each satellite transmits a message containing three pieces of information, the satellite number; it's position in space and the time at which the message was sent. The GPS receiver reads the message and saves the information. GPS receivers take this information and use triangulation (otherwise called 'trilateration') to calculate the user's exact location. The GPS receiver can compare the time at which a signal was transmitted by a satellite with the time it was received by the GPS receiver. This allows it to determine how far away that particular satellite is. With distance measurements from four satellites, the position on the ground can be calculated.

The Figure 4.10(a) shows two known positions, 'Position 1' and 'Position 2' as the dots. The circles show all points at known distances 'd1' and 'd2' from them. These are the data that the GPS receives from the orbiting satellites. There are two possible locations that are the specified distances from the two locations. These are shown by the points where the two circles overlap. The Figure 4.10(b) shows an additional point, 'Point 3' and all positions a known distance 'd3' from it. Let us now see how this additional information allows us to identify a single location on the page.

So in two dimensions, on a plane, three sets of known locations and distances are required to uniquely identify a point in space. In three-dimensional spaces, such as the one in which we live and the satellites orbit, four sets of positions and distances are required to uniquely identify a location. When using GPS, the more that are known from satellite the better to overcome errors in measurement.

A GPS receiver must be locked on to the signal of at least three satellites to calculate a two-dimentional position (latitude and longitude) and track movement. With four or more satellites in view, the receiver can determine the user's three-dimensional position (latitude, longitude and altitude). Once the user's position has been determined, the GPS unit can calculate other information, such as speed, bearing, track, trip distance, distance to destination, sunrise and sunset time and more.

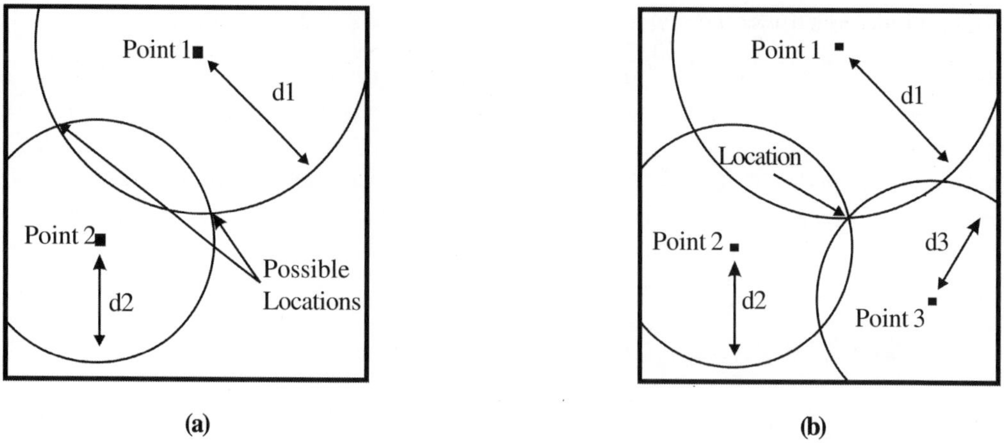

Figure 4.10: Working principles of GPS

4.4.3 Sources of errors in GPS

There are many sources of possible errors that will degrade the accuracy of positions computed by a GPS receiver. Sources that can degrade the GPS signal and thus affect accuracy include the following:

1. Ionosphere and troposphere delays: The satellite signal slows as it passes through the atmosphere. The GPS system uses a built-in model that calculates an average amount of delay to partially correct for this type of error. Sunspot activity also causes interference with GPS signals.
2. Signal multipath: Multipath effects arise when signals transmitted from the satellites bounce off a reflective surface before getting to the receiver antenna. When this happens, the receiver gets the signal in straight line path as well as delayed path (multiple paths). The effect is similar to a ghost or double image on a TV set.
3. Clock errors: A receiver's built-in clock is not as accurate as the atomic clocks onboard the GPS satellites. Therefore, it may have very slight timing errors. Also small variations in the atomic clocks (clock drift) on board the satellites can translate to large position errors; a clock error of 1 nanosecond translates to 1 foot or 0.3 meters user error on the ground.
4. Orbital errors: Also known as ephemeris errors, these are inaccuracies of the satellite's reported location.
5. Number of satellites visible: The more satellites a GPS receiver can "see," the better will be the accuracy. Buildings, terrain, electronic interference, or sometimes even dense foliage can block signal reception, causing position errors or possibly no position reading at all.

GPS units typically will not work indoors, underwater or underground.

6. Satellite geometry/shading: This refers to the relative position of the satellites at any given time. Ideal satellite geometry exists when the satellites are located at wide angles relative to each other. Poor geometry results when the satellites are located in a line or in a tight grouping.
7. Intentional degradation of the satellite signal: Selective Availability (SA) is an intentional degradation of the signal once imposed by the U.S. Department of Defense by introducing artificial clock and ephemeris errors. SA was intended to prevent military adversaries from using the highly accurate GPS signals. When SA was implemented, it was the largest component of GPS error, causing error of up to 100 meters. SA is a component of the Standard Positioning Service (SPS), which was formally implemented on March 25, 1990, and was intended to protect national defense. SA was turned off on May 1, 2000.

4.4.4 Geometric dilution of precision (GDOP)

Satellite geometry can also affect the accuracy of GPS positioning. This effect is called Geometric Dilution of Precision (GDOP). GDOP refers to where the satellites are in relation to one another, and is a measure of the quality of the satellite configuration. It can magnify or lessen other GPS errors. In general, the wider the angle between satellites, the better the measurement. GPS receivers usually report the quality of satellite geometry in terms of Position Dilution of Precision, or PDOP. PDOP refers to horizontal (HDOP) and vertical (VDOP) measurements (latitude, longitude and altitude). One can check the quality of the satellite configuration, the receiver is currently using by looking at the PDOP value. A low DOP indicates a higher probability of accuracy, and a high DOP indicates a lower probability of accuracy. A PDOP of 4 or less is excellent, a PDOP between 5 and 8 is acceptable, and a PDOP of 9 or greater is poor. Another term we often encounter is TDOP, or Time Dilution of Precision. TDOP refers to satellite clock offset. On a GPS receiver we can set a parameter known as the PDOP mask. This will cause the receiver to ignore satellite configurations that have a PDOP higher than the limit one specifies.

4.4.5 Differential GPS (DGPS)

Differential Global Positioning System is used for minimizing the error and to have better positional accuracy through differential corrections. Differential correction is a method used to reduce the effects of atmospheric error and other sources of GPS positioning error (differential correction cannot correct for multipath or receiver error; it counteracts only the errors that are common to both reference and roving receivers). It requires, in addition to "roving" GPS receiver, a GPS receiver on the ground in a known location to act as a static reference point. This type of setup is often called a GPS base station. Since the base station "knows" where it is, it can compute the errors in its position calculations (in reality, it computes timing errors) and apply them to any number of roving receivers in the same general area. This requires that the base and rover receivers "see" the same set of satellites at the same time.

The base station, depending upon how it is configured, can correct roving GPS receiver data in one (or both) of two ways: (1) In the first method, called real-time differential correction or real-time differential GPS (DGPS), the base station transmits (usually via radio link) error correction messages to other GPS receivers in the local area. In this case, the positions you read on your GPS receiver while you are out collecting data, are the corrected positions. (2) The second method, called post-processed differential correction, is performed on a computer *after* the roving receiver data are collected. While you are out in the field collecting data, the positions you read on your roving GPS receiver are uncorrected. It is not until you take your rover files back to the office and process them using differential correction software and data from the base station file, that you get corrected positions. The base station file contains information about the timing errors. This information allows the differential correction software to apply error corrections to the roving receiver file during processing. Since the base and rover receivers have to "see" the same set of satellites at the same time, the base file has to start before the rover file starts, and end after the rover file ends (a base station is normally set up to track all satellites in view, insuring that it will "see" at least the four satellites that the roving receiver is using to compute positions). Post-processed differential correction, then, requires both base and rover receivers that are capable of collecting and storing files. Most recreational grade receivers cannot collect and store files that can be differentially corrected.

4.4.6 GPS Accuracy

The accuracy that can be achieved using GPS depends on the type of equipment used, the time of observation, and the positions of the satellites being used to compute positions on Earth. In general, recreational and mapping grade receivers using C/A code without differential correction are accurate to between 10 and 15 meters. Many people using recreational grade receivers don't realize they cannot get highly accurate readings using them autonomously (without differential correction).

Most mapping and recreational grade receivers with differential correction can provide from about 1 to 5 meter accuracy. Some receivers use what is called "carrier-smoothed code" to increase the accuracy of the C/A code. This involves measuring the distance from the receiver to the satellites by counting the number of waves that carry the C/A code signal. These receivers can achieve 10 cm to 1 meter accuracy with differential correction. Dual frequency survey grade receivers using more advanced network survey techniques can achieve centimeter to millimeter accuracy.

Some people wonder why GPS is better than Loran or other systems that use ground-based transmitters. The accuracy of ground-based location systems such as Loran, which uses low frequency radio signals, is affected by signal distortion, varied terrain, local atmospheric disturbances and limited coverage. Since GPS signals come from satellites, the problems common to ground-based systems can be avoided.

4.4.7 Applications of GPS

More and more producers today are using precision farming techniques that can help increase profits and protect the environment. Precision, or site-specific farming involves applying fertilizer, pesticides and other inputs only where they are needed. GPS-guided equipment is often used for variable rate application of fertilizer (based on soil tests) or pesticides (based on pest survey). GPS can also be used to develop the initial reference maps upon which variable rate applications are based. A GPS system combined with a yield monitor can be used to develop an on-the-go yield map or can be used to map weed locations when harvesting. Mounted in an airplane, GPS can be used to guide aerial spraying operations.

GPS can be used to locate weed, insect or disease infestations and monitor their spread. It can also be used to navigate back to previously mapped infestations to apply controls. A field map can be created using GPS to record the coordinates of field borders, fence lines, canals, pipelines, and point locations such as wells, buildings, and landscape features. The resulting field map might be the first layer a producer would develop for an on-farm GIS. Additional layers showing crop damage from hail or drought, and riparian areas or wetlands could be mapped using GPS. Ranchers could use GPS to develop rangeland utilization maps and to navigate back to previously mapped areas or monitoring sites.

GPS applications in natural resource management include inventory and mapping of soils, vegetation types, threatened and endangered species, lake and stream boundaries and wildlife habitat. GPS has been used to aid in damage assessment after natural disasters such as fires, floods and earthquakes. GPS has also been used to map archaeological sites and for infrastructure (streets, highways and utilities) mapping, management, and planning for future growth. Engineers use GPS for surveying when building roads, bridges and other structures.

Other uses of GPS include real estate valuation and taxation assessment, air quality studies, environmental protection, demographic analysis including marketing studies, atmospheric studies, oil and gas exploration, and scientific exploration. There are many current and possible uses for GPS. Any application where location information is needed is a possible candidate for GPS.

4.4.8 Defining GPS requirement

Before investing in GPS equipment, it is important to clearly define the needs in terms of accuracy level required and the end results expected. Does one simply want to navigate in the woods, or to map out points, lines and areas that can be differentially corrected and imported into a GIS (a computer mapping system)? Is real-time differential GPS required for any reason? Is 15 meter accuracy good enough? If so, one does not have to worry about differential correction. If someone wants to make a map from his/her data, 1-5 meter accuracy may be sufficient; otherwise one has to go for sub-meter accuracy for application. Remember that more accurate the equipment more expensive it is. If high accuracy is needed, be sure to justify the added expense.

In addition, considering the needs for durability and weather resistance, and details such as whether or not an external antenna can be connected to the receiver, and its size, weight and suitability for the method of survey (e.g., will it be used in a backpack, mounted on a vehicle, or carried in your hand?) choice of GPS are to be decided.

Identifying requirements ahead of time will help to determine which type of receiver to purchase, and specific features needed in order to accomplish objectives. It will help to avoid purchasing a receiver which can't perform the way to the expectation. A good strategy is to clearly outline the project requirements and then contact several GPS equipment manufacturers with the specifications. Commonly available GPS instrument in India are Gramin, Trimble, Lieca GS, Magellan etc., cost ranging from rupees 20000 to 15 lakhs.

CHAPTER 5

SPATIAL DATABASE MANAGEMENT

"Everything is related to everything else, but near things are more related than distant things."

Tobler's First Law of Geography

5.1 OVERVIEW

Database design aims to produce a detailed data model of particular database to suit the need of the user for a certain set of applications. The design process roughly involves three stages—planning and analysis for the specific requirement (conceptual designing), logical designing and physical designing. It has two parts, data model and functional model.

A data model is a bridge between the 'concept' of the real world (as one sees it) and the physical representation of the 'concept' i.e., organization of the data into data structures. The data structures include data objects, their associations, and the rules, which govern operations on the objects. While a data model focuses on effective and efficient storing of data, the functional part of the database designing process deals with the processing of the stored database (i.e., the access and operations on the data).

Data modeling is the most intensive and time consuming, as it ensures all data objects required by the database are completely and accurately represented. It should be independent of software and hardware and must be reviewed and verified according to the changes in the user's requirements.

5.2 CONCEPT OF SPATIAL DATABASE

In simpler terms a *database* is a collection of related data. A data without so context has little value. When data are organized, presented, analyzed, interpreted and considered useful for a particular decision problem, they become 'information'. For example, the number '10' is a *data*; but 10 degree Celsius is information, which tells us the 'temperature' of an object. This information resulted from the analysis or organization of some data which creates a database. Thus a database can be used to sort a range of temperature values into order, to calculate the maximum and minimum values and average of temperatures or to convert degree Celsius to degree Fahrenheit. The prompt example of a database may be the telephone directory or a dictionary, where the entries are organized in a structured manner to allow users to have easy access to his queries. However, in this type of database (paper-based), our search options are limited, security of data is almost nil and updating involves reprinting, which is a costly affair. Computer based database removes many of these shortcomings and gives a wide possibilities to record, store, and retrieve, updating, analyzing and even, displaying the data in various formats. The similar is the case of GIS. As the database associated with a map in a GIS environment has essentially the 'spatial' component, it is called the *spatial database*. A spatial database is thus defined as a collection of inter-related geospatial dataset that can handle and maintain a large amount of data, which is shareable between different GIS applications. A computerized (GIS) spatial database is not static, but can be used to model changes in the spatial phenomenon over time. The major functions of a spatial database include (Burrough, 1986):

1. Consistency with little or no redundancy;
2. maintenance of data quality including updating;
3. self descriptive with metadata;
4. high performance by database management system with database language, and
5. security including access control.

In GIS, there are two types of data to be managed, spatial and attribute. An entity has both spatial and attribute data to describe it, e.g., a school is located by the Cartesian (x, y) coordinate (spatial data) and is described by the name, type, color of building etc., (aspatial or attribute data). Conventionally, database store only non-spatial entities; however, all entities in GIS are spatial. All such systems today have in-built or internal database module, which can handle a huge volume of spatial as well as non-spatial data.

5.3 Database Management Systems

Database management systems (DBMS) are the collection of computer programms for organizing and managing the database. They may be very small to run on PCs or large enough to be run only on mainframes. It has different components as:

1. Data types: Integer (whole numbers), real (decimal), character (alphabetic and numeric characters), picture/image data types.
2. Operators: Sort, add, delete, edit, select etc., (of) records.
3. Data definition language (DDL): As the name suggests, DDL establishes and maintains the definition of the database tables, e.g., in SQL (Structured Query Language) the comment CREATE is used to create table definitions.

4. Query language: This is another part of the language used to interrogate and change/manipulate databases (e.g., edit, analysis, input, output etc.).
5. Programming tools: The database management should ensure that the data can be accessed directly from software packages/modules for their subsequent applications.

DBMS make assumption of the data it handles, and these assumptions are necessarily for the effective use of a DBMS. Thus, a DBMS has to be suitable for use in a GIS, where it can handle the spatial data in a better way than others. In many times only the attribute data are accessible through DBMS as the locational data do not fit into the data model. When all the data contained in a GIS can be accessed through DBMS, it is called total DBMS solution. In case of data sharing, the database provides means to separate the data from its application, making the data storage independent to its use. The data can be changed by the programmer/applicant depending on the objective without changing other components of the system.

Out of the four major types of database models, hierarchical, network, relational and object oriented, the relational model has been the most successful in a GIS. Some of the well known relational databases are dBase, Oracle and Info. The object oriented database model is a relatively new concept that is thought to have huge potential in its applications.

5.4 Database structures

It may not usually be essential for a GIS user to know how data are organized in the computer for its optimal storage and access. But some knowledge of data modeling and data structures can give a user the required understanding of how the systems work, and what are their relative advantages and limitations. Thus, before entering into data modelling, let us have a look on various database structures.

Systems for organizing data in the computer range from *simple lists* or *indexed files* at the elementary level to *hierarchical*, *network*, *relational* and *object-oriented* scheme as the highly structured form. A strong database system (structure) provides quick access and cross reference to the data. There are a number of ways to achieve this, but no single 'best' method could be identified. Let us begin with the simple computer file structures.

5.4.1 Simple computer file structure

In the computer language, a 'file' is a collection of related information (records) in a computer that can be accessed by a 'unique' name. A file structure allows the user to store, locate, retrieve and update the records as and when required. The following are some of the major and common computer file structures.

Simple Lists It is the simplest form of a file structure database where any new data is added at the end of the file (like a person enters the bank and stands at the end of the queue at a counter for transaction), which gets longer and longer (the bank employee at the counter is not seen!). As the data is entered without any order, data retrieval becomes very inefficient and time consuming. Thus, for a list containing *n* items, it takes an average of (n + 1)/2 search operations to find a particular record. For a large number of items, for example, if 1 second is required to perform a single record searching, database with 100,000 records may take (100,000+1)/2 seconds or 14 hours. It is, therefore obvious that there have to be some methods, which keep the data in a structured order and provide a 'key' to that structure in order to speed up data retrieval.

Ordered Sequential Files Let us take the example of the telephone directory. Here, unlike the simple listing system, any new entry is placed to its exact position, the position being defined alphabetically. Thus, an extra room is always created for any new entry, and the data retrieval becomes fast and efficient. Here the search method is usually binary. Instead of starting the search at the beginning of the list, the record in the middle is examined first. If the sequence of letters in a word (key value) matches, then the middle record will be the one being sought. If the value does not match, it is checked whether the required item occurs before or after the middle element. If it is before, the upper half is searched. If it is after, the other half is searched in the same manner until the item has been located. This search operation requires log2 (n + 1) steps (a few seconds —hence much faster than the simple list method).

Indexed Files In case of ordered sequential system, data is retrieved using a key attribute (for dictionary, it is the 'spelling'). But in GIS, the individual entities also have many associated attributes. Very often we need to refer to these associated attributes, not just the 'key' attribute like identification number or the name. For example, we have a list of soil profile that has been structured using the key attribute soil series, but quite often we need to have information of the associated attributes like pH, organic carbon, textural class etc. Thus our strategy to access and retrieve data becomes different. As large number of attributes is likely to be associated with the key attribute (or entity like point/area/polygon), we require a more efficient method e.g., some sort of an index to the directory. These indexed files can either be direct or inverted; when the data items in the file themselves provide the main order of the file, these files are known as direct files (Fig. 5.1a). The location of items in the main file may also be specified according to topic, which is given in a second file, known as inverted file (Fig 5.1b). Just as a book, index only can satisfy the preliminary search request by determining the items (pages). In the direct file, the record for each item contains sufficient information for the search to jump over unnecessary items. In the above example, names of the soil series are ordered alphabetically. Each item contains not only the series name but also a number indicating the storage location of that series beginning with the same letter. A simple index file is constructed that lists the correspondence between first letters of series name and storage location. The search proceeds by a sequential search of the index, followed by a sequential search of the appropriate data block. The average number of search steps thus becomes (n_1 + 1)/2 + (n_2 + 1)/2, where n_1 is the number of steps taken in the index and n_2 is the number of items in the data block referenced by the index. But here also, we still have problems with the files in which records are continuously being added or deleted (as often done in interactive mapping systems). In this case not the file only but the index must be modified also.

(a) Index file

Index		File item
Item key	**Record number**	
A	1	A_1
B	n_a+1	A_2
C	n_a+n_b+1	...
...	...	B_1
...	...	...
...	...	C_1

(b) Inverted file

	Attributes					
Soil profile number	**S**	**R**	**P**	**D**	**T**	**E**
1	A	4	Deep	Good	Sandy	No
2	B	5	Shallow	Good	Clay	Yes
3	C	6	Shallow	Poor	Sandy	No
4	D	7	Dccp	Good	Clay	Yes
5	E	4	Deep	Poor	Clay	No
6	F	5	Shallow	Poor	Clay	No

S = soil series, R=pH(Soil reaction), P = depth, D = drainage, T = textural class, E = erosion severity

(c) Index inverted file

Topic	**Soil Profiles (sequential numbers in original file)**					
Acidic pH	1	2	3		5	6
Deep	1			4	5	
Shallow		2	3			6
Good drainage	1	2		4		
Poor drainage			3		5	6
Sandy	1		3			
Clay		2		4	5	6
Eroded		2		4		

Figure 5.1: Simple computer file structure

Flat files and Spreadsheets A flat file or spreadsheet is one of the simplest methods of storing data. All records in this database have the same number of 'fields'. Individual records have different data in each field with one field serving as the key to locate a particular record. For example, the roll number of a Postgraduate student may be the key field in his/her database including name, address, phone number, sex, nationality, place and date of birth, date of admission and so on. For an individual student, there might be hundreds of fields associated with the record. When the number of fields become very lengthy in a flat file, search becomes difficult. Searching by other determinants is also difficult as the key field has been set by the programmer and expanding the number of fields usually necessitates re-programming. Adding new records is also time consuming, particularly when there are numerous fields. Other methods offer more flexibility and responsiveness in GIS.

Hierarchical Files This method is usually described as a "parent-child, one-to-many" relationship (Antennucci et al., 1991; Burrough, 1986; Huxhold, 1991). One field is key to all records, but data in one record need not be repeated in another and records with similar attributes are associated together. A key field links the records in a hierarchy of files. Each record, except for the master record, has a higher-level record file linked by a key field 'pointer'. In other words, one record may lead to another and likewise in a relatively descending pattern. The relationship between the records been clearly defined and queries following a standard routine, hierarchical data structure is very efficient where access to other records becomes readily available through the master record with the key field determinant linking downward to other records.

Relational Files Relational files connect different files or tables (matrices of tables containing information) using a common link, which is not hierarchical (Fig. 5.2). As long as the tables have a common link they may be combined by the user to form new query and data output. This is the most flexible system as on date and is particularly suited to Structured Query Language (Walker et al., 1996; Antennucci et al., 1991; Huxhold, 1991). The flexibility makes it ideal for GIS, mostly because queries are based on relationships from one type of record to another that the user establishes. A comparison of flat, hierarchical and relational files is presented in Table. 5.1.

One must remember that a data model is not the same as a data file. Data models are generic way of working with data whereas a file is manner of storing the actual data so that GIS software can interpret the data values and present them graphically. Following is the discussion of the data models and their data structures.

5.5 DATABASE MODELING

As explained above, data on real world entities are written into computer files, which are simple or organized lists. The basic unit of a file is the data record or *tuple*, which contains all the relevant information for each entity. In all kinds of database structures, data are written in the form of records. The simplest kind of record is a one-dimensional array of fixed length, divided into a number of equal partitions as shown in (Fig 5.3a). This record is ideal when all items have the same number of attributes (e.g., the soil series were sampled and analyzed for a standard range of cations). Here all the data records are of equal length and there is no 'header' (a record that contains information about the type of information in the sub-record and the amount of space it takes up). When the attributes are of variable lengths and the set of attributes are not common for all the series, we have to have variable length records (Fig. 5.3b). Here, every record has a header. Adding information about data location in a header record improves efficiency of data storage.

Database modeling refers to the task of designing a database that performs efficiently, contains correct information, has logical structure and is as easy as possible to maintain and extend

(Worboys et al., 1990). There are different data models, frequently used in GIS as *hierarchical*, *relational*, *network* and *object-oriented* data models. A brief account of these data models are presented below.

Author

Author_id*	Author_surname	Author_firstname	Address	City	State
100					
101					
102					
103					
104					
105					
106					
107					
108					

Author & Title

Author_id*	Title_id*
100	A11
101	A12
102	B34
103	C56
104	E10
104	M25
107	O21
108	S30

Publisher _id**	Publisher_name	City
500		
400		
200		
600		
900		
800		
001		
700		

Title

Title_id*	Title	Type	Price	Publisher_id**
A11				500
A12				400
B33				400
B34				200
C56				600
E10				900
M25				800
O21				001
S30				700

* Primary key column

** Foreign key column

Figure 5.2: A relational database structure

Table 5.1: Comparison between flat, hierarchical and relational files

Structure	**Advantages**	**Disadvantages**
Flat files	Faster data retrieval, simple structure and easy to programme	Difficult to process multiple values of a data item, adding new data categories requires reprogramming and slow data retrieval without the key
Hierarchical files	Adding and deleting records is easy, fast data retrieval through higher level records and multiple associations with like records in	Pointer path restricts access, each association requires repetitive data in other records, pointers different files require large amount of computer storage
Relational files	Easy access and minimal technical training for users, flexibility for unforeseen inquiries, easy modification and addition of new relationships, data, and records, and physical storage of data can be changed without affecting relationships between records	New relations can require considerable processing, sequential access is slow, easy to make logical mistakes due to flexibility of relationships between records

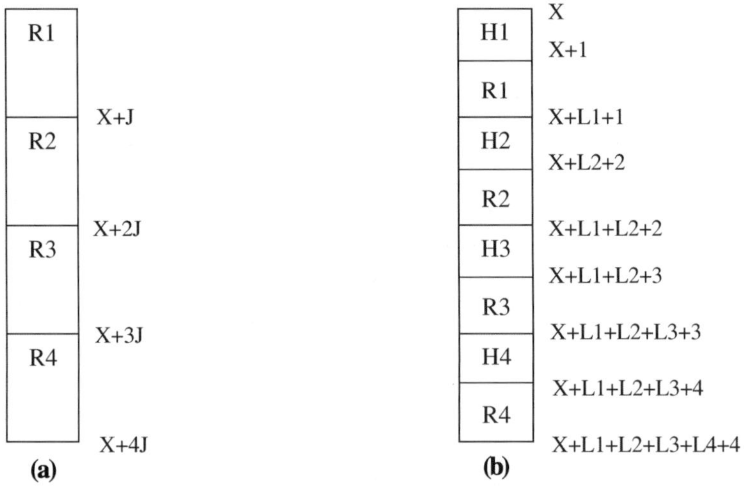

Figure 5.3: Data records

5.5.1 Hierarchical model

A hierarchical data model organizes data in a tree structure. This tree has a set of linked nodes, each node has zero or more 'child' nodes and the tree grows down, not up! The first node has no 'parent' and called the root node or 'root'; each of other nodes has at most one parent, i.e., for every record, there is only one parent record at the next level up in the tree. Node having child is the child's parent node; nodes with no children are called 'leaf' nodes. The links between record types are defined by the parent-child relationships (Fig. 5.4a). Each record contains a 'key' field, which refers to the attribute that assigns records to particular levels in the database structure. Starting from the top, the hierarchy permits a set of elements to be accessed at the next-lower level. Any element may retain connections to only one set of lower-level elements and itself be a member of only one such set. There are no connections between the records at the same level like in network (Fig. 5.4b).

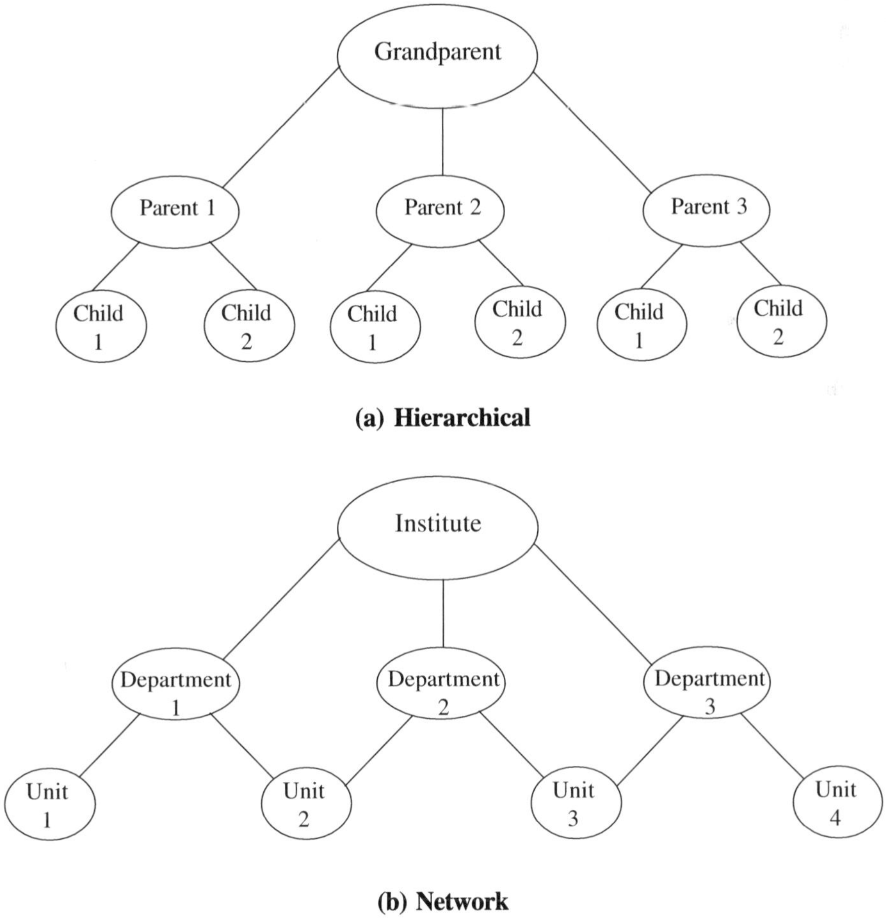

Figure 5.4: Concept of database model

For example, a company or organization might have various departments/ divisions, each with attributes like name of heads, wing numbers, telephone numbers, number of staff, etc. Each department is assigned to carry certain responsibilities in order to collectively make the company's overall success, with attributes related to each of them. Again, each small job may have several components, with their respective attributes, and so on. Certain types of geographical data may fit well with the hierarchical model e.g., census data organized by state, within state by city, within city by census tract etc. These databases keep track of the different record types, their attributes, and the hierarchical relationships between them. Some well-known hierarchical database structures are Adabas, GT.M., IMS, Cache, Multidimensional hierarchical toolkit, mumps compiler etc.

Let us now see how hierarchical model can be used to store spatial data in a GIS. In a vector data structure the spatial data can be stored through the hierarchical structure as explained in Figure 5.5. There are two polygons (I, II) in the map. Each of these two polygons consists of few lines and the lines, in turn, are made up with points. Altogether there are 12 lines to compose the polygons, and 11 points at the intersection of the lines. Thus the map is at the top of the hierarchy and points are at the bottom.

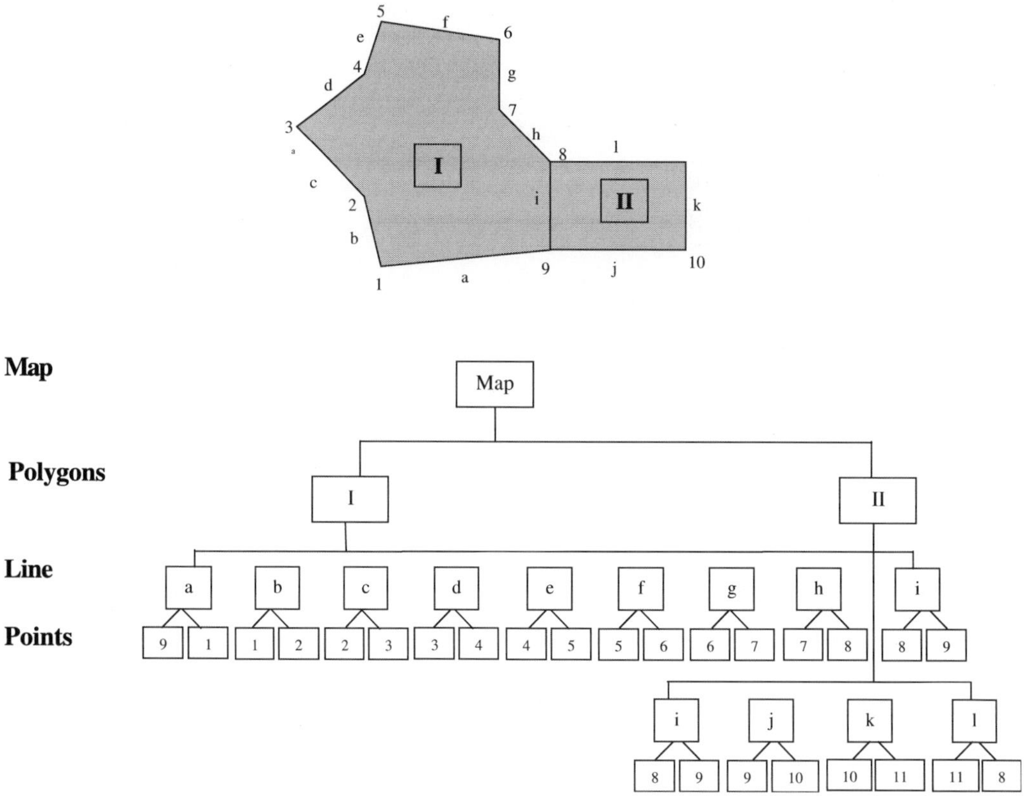

Figure 5.5: Spatial data storage in the hierarchy database model

In a raster model, different data layers are stored in such a way that a layer at any level consists of the algebraic average of data of its immediate lower level (e.g., a layer of data with a resolution to the degree of 10 mega pixel size can be re-sampled to a layer having resolution of 40 mega pixel size). In this process we can visualize various data layers stacked together like a pyramid. Hence, hierarchical data structure can also be called pyramidal data structure (Star and Estes, 1999). The *quadtree* used to access a small part of a large raster image or map area, is a type of hierarchical model (more detailed discussion is given in section 5.6.4).

Hierarchical model is easy to understand and it has relatively high speed of access. It has also ease of updating and expanding the existing datasets. The 'tree' structure is very common for natural objects (always found in taxonomic classification), and it is a rigid structure. The linkages are only through one-to-many connections and the search routines are fixed and constrained. New links or new search routines cannot be defined without restructuring or creation of new keys. This structure implies that a record can have repeating information, generally in the child data segments; hence, data redundancy is one of its biggest disadvantages. Unusually large index files are maintained; attribute files are repeated taking unnecessary memory space. These constraints impose the presupposition that all queries can be known in advance and accounted for structuring and entering data, something which is not necessarily either natural or suitable in GIS applications. As a result, hierarchical database systems are usually restricted to storing digital map data in GIS, a very effective solution where the data structure is naturally hierarchical. Even at times, a hierarchical data structure may refuse to answer to certain queries, as the information needed is not contained in the structure (Burrough, 1983).

5.5.2 Network model

A hierarchical structure fails if one-to-many relationship is violated (e.g., a child can have more than one parent) then the hierarchy becomes a network data structure. In case of geographic features, the adjacent items in a map need to be linked together. Networking structure is the most suitable here (Fig 5.4 b). In real world, data has to be modeled with more than one parent per child. In the network structure, each data element has connections to several different-level elements. This network structure represents the complex interrelationships (one-many, many-one, and many-many) that often exist between real-world geographical objects. Unlike 'tree' structure, network model has 'lattice' structure.

This type of data model was developed by Charles W. Bachman, an outstanding computer/database scientist, as a part of work of CODASYL (Conference on Data Systems Languages), which proposed programming language COBOL (1966). In 1971, CODASYL formally defined and developed the network model with the objective to reduce the multiple entries of the hierarchical structure (data redundancy), thereby, separating data structure from physical storage. This model allows a more natural modeling among entities. The cost and errors could also be largely eliminated as points and lines are entered only once (Fig. 5.6). The duplicity as seen in hierarchical data model is eliminated by establishing multiple connections among the elements at different levels. Hence, the points are repeated, making entry of a number of points in hierarchical structure; while in this model, only 11 points are required (Fig. 5.5 and Fig. 5.6). The common boundary between polygon I and II (line 'i') need not be printed every time it is referenced by the polygons, thus making the map generalization easier.

Based on mathematical set theory, network model defines one-many relationships with a set consisting of an owner record type, a set name, and a member record type. The member record type can be member of more than one set, thus multi-parent concept is established. Similar is the case for owner record type. Several pair of sets forms the complete network—in each set one record type is owner and one (or more) record types are members. Thus, the model has more flexibility than the hierarchical type for handling complex spatial relationships. But because of IBM's choice in sticking to its IMS and DL/L database and information management systems, which are hierarchical in nature and also the introduction of more productive relational data model vis-à-vis availability of faster hardware options, it could not become dominant and hence is less frequently used in GIS applications.

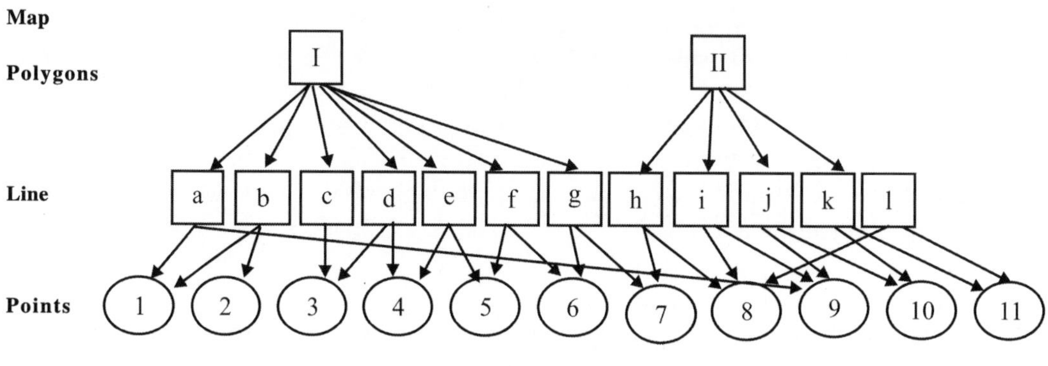

Figure 5.6: Spatial data storage in a network model

5.5.3 Relational model

Relational model proposed by Dr. Edgar Frank Codd, the famous computer scientist in 1970 and subsequently improved by Chris Date and Hugh Darwen is so far the most popular model for GIS. The relational model assumes that all data are represented as mathematical relationships where only two possible evaluations exist for each preposition—true or false. The operation of data is performed by using relational calculus and algebra. INFO in ARC/INFO, DBASE III for several PC-based GIS and ORACLE for several GIS packages are built on relational approach. Understanding the basics of a relational model is important to effectively use relational database like Oracle, Microsoft SQL, or even the personal database systems like Access or FOX. Data structures, relationships, and data integrity are the basis of the relational model.

5.5.3.1 Basics of relational data model

The fundamental understanding of the relational model starts with database as a collection of relational tables (Codd, 1970, 1982). A relational table is a flat file composed of a set of named columns and an arbitrary number of unnamed rows. A data value is stored in the intersection of a row and column as shown in Fig. 5.2. Each row and column in a table is called a 'record' or 'entity' or 'tuple' and a 'field' or an 'attribute', respectively. Thus each record consists of a set of attributes while each table (visual representation of 'relation') consists of different number of attributes, which is called the *degree*. The degree of an attribute refers to *n*-ary (unary, binary etc.) relation. Table with one attribute is a unary relation (name of educational institutes), with two attributes as binary (departments and courses taught), with three attributes as trinary (name, roll number and addresses of the students enrolled) and with *n* attributes, n-ary.

A relational table has six properties (Information Technology Services, 2004) as,

1. *Values are atomic:* This is very important keystone of a relational table where columns are not repeating set of data arrays (such tables are in the "first normal form", 1NF). This simplifies data manipulation logic.
2. *Different kinds of column values:* This means that all values in a column belong to a single domain e.g., a weekly averaged rainfall column must contain only the weekly rainfall of different locations ruling out the mixture of other information like deviations from normal value, or daily rainfall data. This property simplifies data access as well as data validation. Because all values are from the same domain, the domain can be defined and enforced with the Data Definition Language (DDL) of the database software.
3. *Each row is unique:* This property ensures that no two rows in a relational table are identical. There must be at least one column, the values of which uniquely differentiate and identify each row in the table. Such columns are called primary keys (Fig. 5.2). The relation using primary and foreign keys is discussed in more detail in Box. 5.1.
4. *The columns have no proper sequence:* The ordering of the columns in the relational table is insignificant as they can be retrieved in any order and in various sequences. The benefit of this property is that it enables many users to share the same table without concern of how the table is organized. It also permits the physical structure of the database to change without affecting the relational tables.
5. *The sequence of rows is insignificant:* This property is analogous to the one above but applies to rows instead of columns. The main benefit is that the rows of a relational table can be retrieved in different order and sequences. Adding information to a relational table is simplified and does not affect existing queries.
6. *Each column has a unique name:* Because the sequence of columns is insignificant, columns must be referenced by name and not by position. In general, a column name need not be unique within an entire database but only within the table to which it belongs.

Integrity of data is the prime feature of a relational table, which ensures correct and consistent navigation and manipulation of the tables in the database. There are two basic rules to ensure data integrity - entity integrity and referential integrity. The entity integrity rule states that the value of the primary key can never be a null value (a null value is one that has no apparent value and is not the same as a blank). Because a primary key is used to identify a unique row in a relational table, its value must always be specified and should never be unknown. Insert, update, and delete operations must maintain the uniqueness and existence of all primary keys. The referential integrity rule states that if a relational table has a foreign key, then every value of the foreign key must either be null or match the values in the relational table in which that foreign key is a primary key.

Box 5.1 Relationships, attributes and keys

Relationship is an association between two or more tables.

Attributes are data items that describe an entity. For example, student ID, name and academic year are attributes of the entity 'student'. M-3984, Sunil Saxena and M.Sc. II year are the *attribute instance* of 'student'.

The *primary key* is an attribute or a set of attributes that uniquely identifies a specific instance of an entity. Every entity in the data model must have a primary key whose values uniquely identify instances of the entity. In a relational table, primary key uniquely identify each row in a table and must have a non-null value. When an entity is identified by more than one attribute, the key is called *candidate key*. In the present example, there may be two candidate key, student ID and name; if student ID is chosen as primary key, the name is also termed as *alternate key*.

A *foreign key* is an attribute that completes the association between the child and the parent entity. It is like a copy of the primary key from another relational table. The relationship is established between two relational tables by matching the values of the foreign key in one table with the values of the primary key in another.

Keys are fundamental to the concept of relational database because they enable tables in the database to be related with each other.

Source: http://www.utexas.edu/its/windows/database/datamodeling/rm/rm4.html

The organization of data into relational tables is known as the 'logical' view of the database. It is the form in which a relational database presents data to the user and the programmer. The way the database software physically stores the data on a computer disk system is called the 'internal' view. The internal view differs from product to product and is of generally no concern to a GIS user as the internal architecture varies substantially from one relational DBMS (RDBMS) to another. This approach is more of a concept than a data structure. In a relational model, the following two important concepts should be defined. One is that there is no hierarchy and the second, no specific keys point to the various records. All fields in a line may be regarded as being connected (or jointed) to each other, and any field in a table may be a key for accessing data from another table. The system permits all objects and attributes to be related to each other. Thus a collection of data on many objects with complex relationships may result in a large number of tables. This is a weakness of the relational approach. The key of relation refers to the unique identification e.g., the key attributes is a telephone directory in a set of last name, first name and address. Any key attribute selected and tabulated should keep the key's uniqueness (non-redundancy) e.g., address cannot be dropped from telephone address, because there may be many with the same names. Prime attribute is an attribute listed in at least one key. The most important point of the relational database design is to build a set of key attributes with a prime attribute, so as to allow dependence between attributes as well as to avoid loss of general information when records are inserted or deleted.

A relational database has three basic attributes or functions: primary key, relational join, and normal forms. Values in a single column, or a combination of values in multiple columns can be used to define a primary key for the table and act as the addressing mechanism as explained above. A function called relational join links data that lie in different tables. Normal form is the identification of interdependence between the data. There are no pointers in the data tables. Internal index tables are used to direct inferable communications, which means that the system must open at least one index table for each connection between data tables. This can result in relatively large databases and slow access. Moreover, as there are no pointers, searches for records or fields must be sequential through the tables. The result is that relational database system is inherently slower than hierarchical or network database systems. However, a field may be located quickly in a given table, and the data accessed may be manipulated.

5.5.3.2 Relational data manipulation

If we observe closely, we see that relational tables are all mathematical sets, where each row of the table is an element of the set. Any operations performed on sets can well be applied on relational tables (Fig 5.7). Accordingly there are eight operations as: union, subtraction (differentiation), multiplication (product), division, intersection, projection, selection and joining. The union operation of two relational tables is formed by appending rows from one table to those of a second table to produce a third. Duplicate rows are eliminated. The difference of two relational tables is a third that contains those rows that occur in the first table but not in the second. Naturally the subtraction from a set 'A' from another set 'B' is not the same as 'B' from 'A'. The intersection of two relational tables is a third table that contains common rows. Union, subtraction and intersection require that the two tables must be compatible i.e., the number of rows and columns must be the same in both of the tables. The product of two relational tables, also called the Cartesian product, is the concatenation of every row in one table with every row in the second. This is an intermediate process of joining operation. Joining operation also combines selection and projection. If the joining criterion is based on equality of column value, the result is called an 'equijoin'. A natural join is an equijoin with redundant columns removed. The division operator results in column values in one table for which there are other matching column values corresponding to every row in another table.

k	x	y
1	A	2
2	B	4
3	C	6

A

k	x	y
1	A	2
4	D	8
5	E	10

B

k	x	y
1	A	2
2	B	4
3	C	6
4	D	8
5	E	10

A union B

k	x	y
2	B	4
3	C	6

A - B

k	x	y
4	D	8
5	E	10

B - A

k	x	y
1	A	2

A intersect B

k	x	y
1	A	2
2	B	4
3	C	6
4	D	8
5	E	10

D

k	z
1	20
4	24
5	28
7	32
9	36

E

k	x	y	k	z
1	A	2	1	20
4	D	8	4	32
5	E	10	5	36

Equijoin

k	x	y	z
1	A	2	20
4	D	8	32
5	E	10	36

Natural join

k	x	y
10	1101	A
10	1201	B
10	1301	C
20	1201	B
30	1101	A
30	1201	B
30	1301	C

A

k	x
1001	A
1201	B
1301	C

B (Divisor)

k
10
30

Result

Figure 5.7: Manipulation involving two relational tables

5.5.3.3 Normalization

Normalization is a design technique that is widely used as a guide in designing relational databases. It is essentially a two-step process that puts data into tabular form by removing repeating groups and then removes duplicated data from the relational tables.

Normalization theory is based on the concept of normal forms. Each norm illustrates some 'rules'. A relational table is said to be a particular normal form if it satisfies a certain set of constraints. There are currently five normal forms of which the first three have been defined by E. F. Codd. The highest (fifth) form implies a well-structured database. The goal of normalization is to create a set of relational tables that are free of redundant data and can be consistently and correctly modified. This means that all tables in a relational database should be at least in the third normal form, 3NF (even the fifth normal form is not a perfect one, but much improved). A relational table is in 3NF if and only if all non-key columns are (a) mutually independent and (b) fully dependent upon the primary key. Mutual independence means that no non-key column is dependent upon any combination of the other columns. The first two normal forms are intermediate steps to achieve the goal of having all tables in 3NF. After third normalization, all normalization problems involve only tables, which have three or more columns, and all the columns are keys. Many practitioners argue that placing entities in 3NF is generally sufficient because it is rare that entities that are in 3NF are not also in 4NF and 5NF. They further argue that the benefits gained from transforming entities into 4NF and 5NF are so less that it is not worth the effort.

Simply stated, normalization is the process of removing redundant data from relational tables by decomposing (splitting) a relational table into smaller tables by projection. The goal is to have only primary keys on the left hand side of a functional dependency. In order to be correct, decomposition must be compact. That is, the new tables can be recombined by a natural join to recreate the original table without creating any spurious or redundant data.

Normalization is concerned with finding the simplest structure for a given set of data. It deals with dependence between attributes and avoids loss of general information when records are inserted or deleted. Although normalization produces a consistent and logical structure, it has a cost in increased storage requirements. Some GIS database administrators avoid full normalization for this reason. Table 5.2 shows how to build the relational database by normalizing an unstructured table.

Codd's 12 rules (Codd, 1970) define the relational database systems. Today, almost all RDMBS employ SQL for their query language, but according to Codd's rules, these databases are not actually RDBMS. Database systems that follow some but not all of the rules have been named as *Pseudo-Relational Database Management Systems* (PRDBMS) in contrast to *Truly-Relational Database Management Systems* (TRDBMS) that do follow all of the rules.

Table 5.2: Normalization to relational database (Unnormalized)

(a) Unnormalized table

Society member	Type of profession	Monthly salary (Rs)
Bose	Doctor	20,000
Chakraborty	Teacher	12,000
Kumar	Doctor	20,000
Lal	Farmer	9,000
Mukherjee	Doctor	20,000
Sahoo	Teacher	12,000
Thakur	Farmer	9,000
Tyagi	Teacher	12,000

(b) Normalized tables

Society member	Type pf profession
Bose	Doctor
Chakraborty	Teacher
Kumar	Doctor
Lal	Farmer
Mukherjee	Doctor
Sahoo	Teacher
Thakur	Farmer
Tyagi	Teacher

Type of profession	Monthly salary (Rs)
Doctor	20,000
Teacher	12,000
Farmer	9,000

A relational model is based on formal, very strong, proven mathematical theories. But in most of the cases it is static, though dynamic relational database systems are also developed. RDBMS is used most frequently for attribute data in GIS applications, primarily because of their simple, flexible structures. Another reason may be that they support the complex relationships common among real-world geographical objects. The structured query language (SQL) is not suited to geometric data but are quite effective for attribute data. Commercial RDBMS have been developed specifically for handling geographical data (and binary large objects for handling of multimedia data). These are new solutions, which so far had very little applications. The most usual database solution for GIS is a combination of file-based systems for geometrical data and standard relational DBMS for attribute data.

5.5.4 Object oriented data model

There are many object-oriented database (OODB) systems existing today, but no consensus could be made to define these systems or to converge them into class or family. However, OODB systems are receiving a lot of attention from both experimental and theoretical standpoints, possibly because the hierarchical, network, and relational database fail to actually represent the conceptual data model of real world system. Moreover, spatial and attribute data are often considered as separate database. Object-oriented database systems are more representative of the real world, where complex relationships exist between the entities. An object is a model of an entity, which has a state (the set of values for the attributes of the object) and a behaviour (the set of rules or methods or program codes, which operate on the state of the object). The objects having the same set of attributes and methods are grouped as a class. An object must belong to only one class. There can be subclass derived from a class, which has all the attributes and methods of the respective class plus some additional attributes and methods.

There are many features for distinguishing the traditional database system from OODBS, some of them are listed below:

a. *Generic Properties:* There should be an inheritance relationship with many number of classes and subclasses.

b. *Abstraction:* Objects, classes and super classes (parent of subclass or classes) are to be generated by classification, generalization, association and aggregation.

c. *Adhoc queries:* There should be simple yet efficient and application independent ways to satisfy the queries. Users can have any spatial operation to obtain spatial relationships of geographic objects using some concise 'high level' languages.

Let us try to represent a thought in GIS: "Andaman is a group of islands and is a part of India". In this case, we do not mind the geographic location with latitude and longitude in the conventional GIS model. It is not appropriate to use the layers. In an object oriented model, we are more careful with spatial relationships, for example, "is a" (island is a land) and "part of" (it is a part of India). In addition, Andaman has Port Blair city and is also in Bay of Bengal region. An object-oriented database is based on a semantic model as shown in Figure 5.8, which is usually managed by a special language.

The main advantage of object-oriented database system is that they are more efficient to manage the complex relationships between objects through an easy maintenance and updating. For example, a line may be a boundary of land or property, a road edge, or something else. In an object-oriented database, it needs to be updated only once, while in other database systems the data pertaining to it may be in various locations, requiring multiple updating operations. At present, object-oriented database systems are not used widely in GIS (especially for commercial); a few GIS use them in manipulating data, because of their ability to handle lengthy transactions in a multi-user distributed environment.

Apart from the above-mentioned models, there are other alternatives like deductive database approach. It allows more complex objects and modeling to be constructed that may be undertaken using relational, network or hierarchical structures and so offer great potential for spatial data handling. But this is still at the theoretical stage.

In the history of the database system, it is found that the paradigms have shifted from the hierarchical to network, relational, and then finally object-oriented models. One paradigm replaces another one in such a way that the problems caused by the previous paradigm can be solved by the new paradigm.

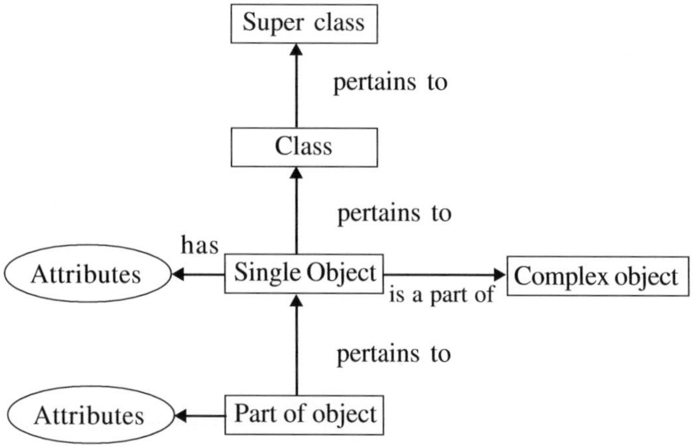

Figure 5.8: Scheme of an object oriented data model

5.6 Data Organization

The database structures as listed and discussed above determine how spatial data can be stored in a computer. The problem with raster data is their size, as some value must be stored for each pixel in an image. An area having a single identity like a lake or a forest requires same storage space as the equal area with many identities (like soil map). Hence the raster data structure takes a total of n rows $\times$ n columns to encode each overlay, in addition to other general information like data type, grid size, projection, etc. Thus, there are two main points when considering storage of raster data within a single file or file structure (a) Space – for example, the available computer storage (b) Speed – the required response time for the application users. In general we do not have control over the storage methods available as these depend on the GIS software capabilities, but if we are designing and implementing a new system or are involved in complex scientific modeling, storage methods will be important to consider. There are four general methods by which the large spatial data can be stored most economically utilizing a minimum memory space. These are (a) chain codes, (b) run-length codes, (c) block codes and (d) quadtrees.

5.6.1 CHAIN CODING

This method defines the boundary of the entity as a sequence of unit cells starting from and coming back to the origin. The cardinal direction of travel around the boundary is usually given using a numbering system (say, 0 = north, 1 = east, 2 = south and 3 = west). For example, the boundary cells of a region can be coded as (start from cell 2 and column 4) N,2 E,4 S,1 W,1 S,1 W,1 S,2 W,1 S,1 W,1 N,1 W,1 (Fig 5.9a). The letters (N, E, W, S) are used here for clarity, but the computer stores them as numbers. Chain codes provide a very compact way of storing data and are good for area analysis and raster/vector transform. Certain operations like estimation of area and perimeter or detection of sharp turns are easy. But this method, sometimes called *Freeman chain coding,* is not good for union and intersection analyses. Redundancy of data also occurs as all boundaries between regions must be stored twice (Freeman, 1974). The greatest advantage of chain coding is reduction in storage volume and the disadvantage is loss of generality. Variations on this technique are particularly applicable for data obtained from scanners and digitizers as these devices have built-in resolution. Scanners have a resolution equal to the pixel size of the scanner, and all displacements between points are actually whole numbers of pixels. Digitizers also work in discrete units.

5.6.2 RUN-LENGTH CODING

Run-length codes reduce the data volume by allowing the points to be stored as row-to-row basis. This method exploits the fact that many datasets have large homogeneous regions. In this procedure, adjacent cells along a row that have the same value are treated as a group and termed a *run* (Fig. 5.9b). Each row in the grid is examined in turn, and pixels having the same value, i.e., homogeneous pixels are grouped together. It starts with left-most cell in each row, and count numbers of like values. For the same area the run-length codes would be as follows

10, 10, 1		
0,10		
0,3	1,5	0,2
0,3	1,6	0,1
0,4	1,5	0,1
0,5	1,3	0,2
0,5	1,3	0,2
0,5	1,2	0,3
0,10		
0,10		
0,10		

The first line in the above represents the dimensions of the matrix (10×10) and the number of entities present (1). In the second and subsequent lines of the file first line of the pair (1 or 0) indicates the presence and absence of the area in the map.

In run-length encoded data, the original data are now replaced by pairs or *tuples*. The first digit in the pair refers to a counter indicating the number of repetitions of the second digit, the data value. So if we had data values of the following order in each of the cells of an original data matrix, in run length encoded form we would have a much more compressed version, viz. original data values in each cell

1 1 1 3 3 3 2 3 3 3 3 3 3 1 1 Total = 15

Run length encoded tuples

(3 1) (3 3) (1 2) (6 3) (2 1) Total = 5

The original data set of 15 elements can now be compressed to 5 elements for a compression factor of about 67 per cent. There are two ways in run-length coding system to store data.

5.6.2.1 Standard run-length coding

In this method the row number, value of the attribute and the number of cells in the run are recorded. This is referred to as a standard method because the row number is always preserved in the dataset, so that the original dataset may be reconstructed if the need arose. As an example the following show an original dataset of 60 elements and the run encoded version that has reduced the number of elements by 50 per cent.

Original data values in each cell

A A A A A A A A A A A A A A A
A A A A A A A C C C C C C C C
A A A B B B B C C C C C A A A
B B B B B B B C C C A A A A A

Here total number of elements = 60. The standard run length encoding will be:

Row	Value	Length
1	A	15
2	A	7
2	C	8
3	A	3
3	B	4
3	C	5
3	A	3
4	B	7
4	C	3
4	A	3

where total number of elements is reduced to 30.

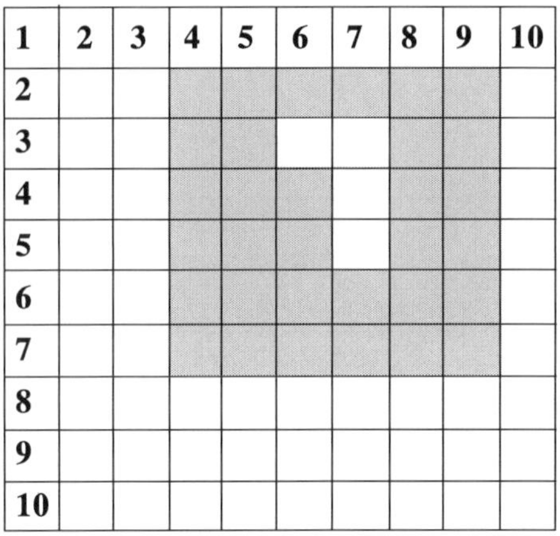

(a) Chain coding

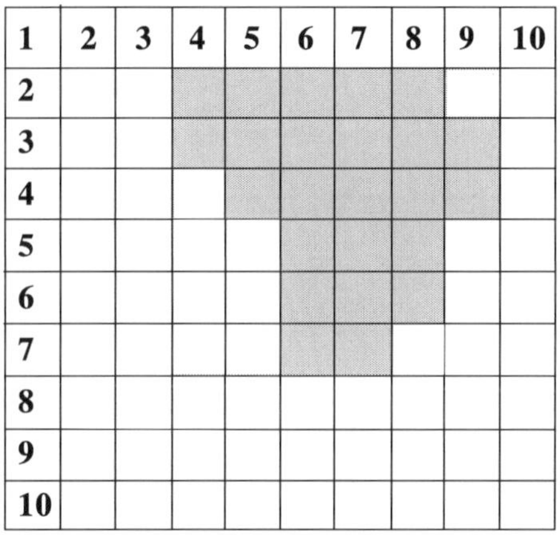

(b) Run-length coding

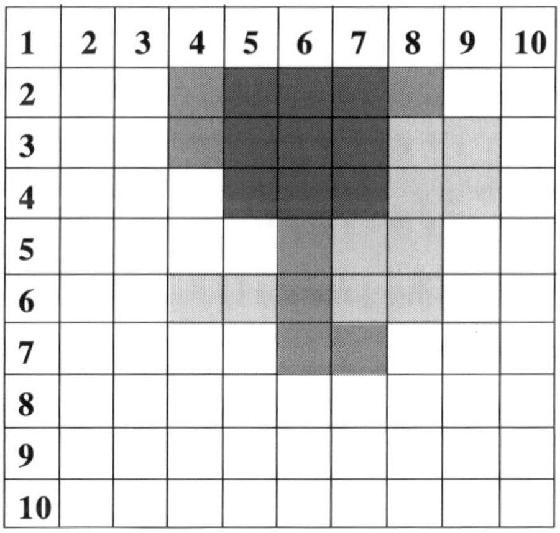

(c) Block coding

Figure 5.9: Compression methods of raster data

5.6.2.2 Value point encoding

Here the procedure begins with assigning position numbers to each cell starting at the upper left corner and proceeding left to right and then from top to bottom of the cell matrix. The position number for the end of each run is recorded in a point column, while the value for each cell in the run is in the value column. With this method we will be able to identify the start and end of cells of a value by simply looking down the table. From an original dataset of values given, it will be seen that the original 60 elements can now be reduced to 18 using the value point encoding method (a 70 per cent reduction).

Original data values in each cell

A A A A A A A A A A A A A A A
A A A A A A A C C C C C C C C
A A A B B B B C C C C C A A A
B B B B B B B C C C A A A A A

Total number of elements = 60; after encoding, total number is reduced to 18 only.

Value	*Point*
A	22
C	30
A	33
B	37
C	42
A	45
B	52
C	55
A	60

Run-length coding is the simplest storage method with considerable improvement in storage requirements. A disadvantage of this method is that pixel groups are identified only in one direction (parallel to the x-axis) so that nearby rows, i.e., those above and below the current row, which may have the same values of the group, are represented separately. Thus, it is not suitable for coding continuous variation. Another disadvantage may be the increased requirements during cartographic processing and manipulation if the data compression is too much.

5.6.3 Block coding

Block codes are the two-dimensional extension to run-length codes using a series of square blocks to tile the area to be mapped (Burrough, 1986). The data structure contains the origin square and the radius of each square. This is called a medial axis transformation (MAT) (Rosenfeld, 1980). The aim is to obtain as many large square blocks as possible for representing data values of the same type. Thus, the larger the square that may be fitted into a region, the more efficient the block coding becomes. The file structure for the block-coded data will be as follows:

Block size	**Number**	**Cell coordinates**
1	7	4,2 8,2 4,3 6,5 6,6 6,7 7,7
4	2	8,3 7,5
9	1	5,2

Therefore ten data blocks are required to store data for the area. These are seven four-unit cells, two four-cell squares and one nine-cell square. Here the top left-hand cell in a block is used as the locational reference for the block (Fig. 5.9c). This method has the advantage of performing union and intersection of regions and for detecting properties such as elongation (Rosenfeld, 1980).

5.6.4 QUADTREE

A quadtree is a term used to describe a hierarchical or tree-based data structure whose common property is that the structure is based on the principle of the recursive decomposition of space. Quadtrees are differentiated by the type of data as used to represent, the principles governing the decomposition process, and the type of resolution, whether variable or fixed. In a quadtree, each *node* in the tree can be represented by two items of information, defining whether or not it contains further subdivisions. If there are no further subdivisions it is a *leaf* in the tree and stops there, and assumes the attribute value associated with it. The quadtree structure can be used for point data, for regions, surfaces and volumes. However, its widespread adoption is hampered by the fact that it is not well suited for resource management applications. Moreover, as it is very recent in usage as a data structure, knowledge and expertise are lacking even though it has been used previously in map organizations for the storage of map data.

The quadtree structure can best be described as the family hierarchical structure like

1. A quadtree has a root (grandfather/grandmother).
2. The root gives rise to four leaves (father/mother).
3. One leaf has four nodes (son/daughter).
4. One of the nodes has four branches (future offsprings). Figure 5.10 depicts a simple quadtree structure.

Raster data has the advantage that each cell can be subdivided into smaller cells of same shape and orientation (Peuquet, 1990). Our area of interest can similarly be divided into blocks and the data can be stored as quadtree (Fig. 5.10). Quadtree addresses both the resolution and the redundancy issues directly. The quadtree seems to be a more compact version for data storage and has the advantage of using a variable sized grid cell. Thus, instead of dividing up an area into cells of one size, finer and finer subdivisions are used only in those regions where finer details have been recorded. These regions would include those data showing points, lines and polygon boundaries where a transition or change from one data value to another takes place. The conceptual framework of the quadtree rests on a process of subdivision based on specific criterion. Here the best resolution that may be obtained is governed by the cell size of the original dataset. Operations such as point-in-polygon searches and the whole family of set-theoretic operations can be performed on quadtree-structured data. It also means that it is easy to generate data to any level of detail. However, a major disadvantage is that it is very time consuming to create a quadtree data structure. This implies that changes to the original data will require a re-building of the entire quadtree. The biggest problem associated with quadtrees is that "tree translation is not translation-invariant i.e., two regions of the same shape and size may have quite different quadtrees, so consequently shape analysis and pattern recognition are not straight forward" (Burrough & Mc. donell, 1998). So, translations of the map or dataset by either rotation or scaling can pose considerable problems with a quadtree data structure. It is not easy to modify a quadtree once it has been built, so more complex maps pose greater challenges. The quadtree is the best suited for maps or datasets containing large homogeneous regions. A region may be split up into parts and a region can also contain 'holes'.

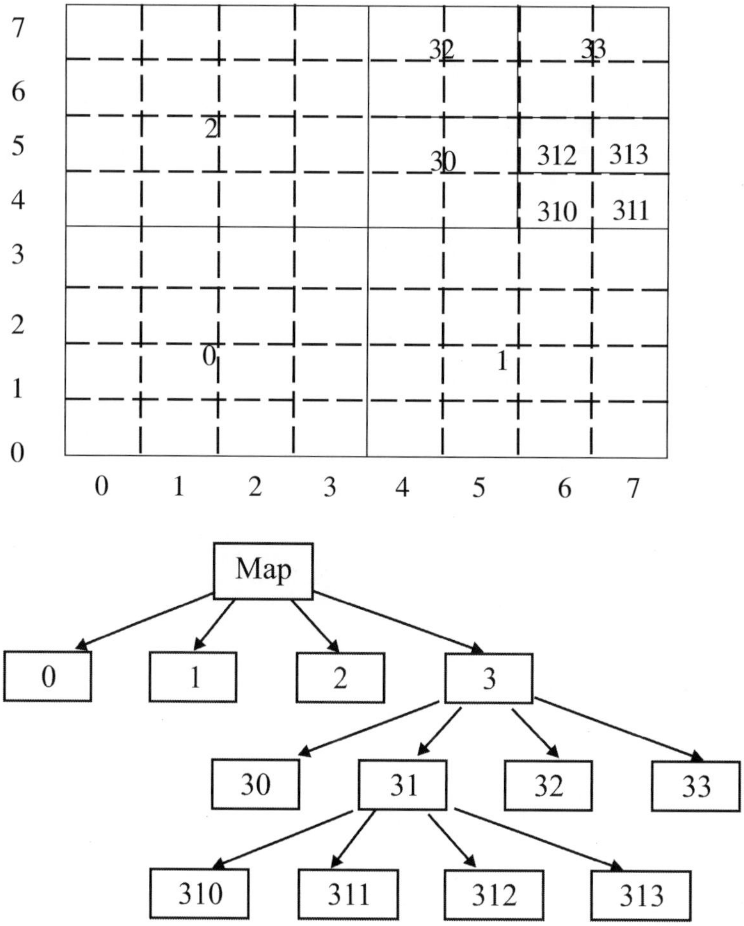

Figure 5.10: Quadtree structure

CHAPTER 6

SPATIAL ANALYSIS

"GIS are simultaneously the telescope, the microscope, the computer, and the xerox machine of regional analysis and synthesis of spatial data"

Alber, 1988

6.1 OVERVIEW

Let us start with an example. The query "How many bore-wells are in Delhi?" is 'aspatial' in nature, as it only answers the number, and we need not to find out where of these wells are located. But if the query goes like this – "How many bore wells are in Delhi within 50 km from the river Jamuna, more than 100 m in depth and can yield 1000 Cusec of water in a day?" the answer is definitely not as easy as the first one because it is 'spatial' in nature. One has to physically locate the wells and find out the required answers to the queries. This explains what we call as the *spatial relationship*. A GIS can readily provide both kinds of information as it has the ability to link spatial data (location) with information (facts and figures) about a particular feature on a map. The information is stored as attributes. Mere visual interpretation of a street network represented by simple lines will not yield much information. To obtain more information like type (metallic or *kaccha* or gravelly) or length or number of sharp turns, one has to take the street name from the map and use it to query a separate database (attributes). GIS allows illustrating the database on the map by using different symbols to draw the types of roads such as by drawing heavy line, regular line, or dotted line. The result of such a display is that the user can determine information about features on the map simply by looking at them. All of these can be done using spatial analysis function, a unique feature that distinguishes a GIS than any other information system.

Spatial analysis is "a set of techniques for analyzing spatial data". It is an umbrella term covering many different procedures as performed using a GIS platform. Although this Chapter is about 'spatial analysis', an eleborate treatment of the topic is beyond the scope of this book. GIS and spatial analysis are one or the same and the whole book is devoted to this. Nevertheless, this Chapter introduces some major spatial analysis functions available in GIS with their possible applications. These functions are based from inductive to deductive approach and can be highly technical, mathematical or very simple and intuitive.

Spatial analysus can be defined as the analysis of the spatial (geographical) data and their attributes for 'decision support'. The cost of input to a GIS is high, and can only be justified by the benefits of analysis/modeling/decision-making performed with the data. The decision or policy makers require quality information at various levels of details depending on their objectives. Geographical analysis transforms the raw data into useful information that can satisfy these requirements. In this process decision models are used which seek to answer different alternatives in solving a problem. Spatial analysis ranges from simple arithmetic and logical operations to complicated model analyses. These analyses are effective not only in the context to which they are used, but also to predict phenomena at other time and place. An inherent problem with geospatial analyses is the difficulty to measure the spatial relationships as the data vary widely in form and compatibility and uncertainty due to scale. As the objectives vary, the analyses become difficult to choose. GIS tries to address many such problems. The list of possibilities to execute analyses using the GIS is endless (ESRI's Arc/ Info GIS has more than 1000 commands or functions). A list of about sixty basic data manipulation and analysis functions can be found in Goodchild (2001).

Spatial function capabilities of GIS were grouped in different ways as re-classification, overlay, connectivity and neighborhood analysis (Berry, 1987); focal, local, zonal and global (Tomlin, 1990); capture, transfer, edit, restructure, transform, query, presentation, etc. (Maguire, 1991); query, transform, measurement, descriptive summary, optimization and hypothesis testing (Goodchild, 1988; 2001); and so on. In this section, we will categorize the spatial functions available in a full GIS as follows:

1. Measurement
2. Query
3. Attribute based operation
4. Neighborhood analysis
5. Connectivity analysis
6. Overlay operation
7. Coverage rebuilding

Each of these is briefly described here under different heads.

6.2 Measurements in GIS

This is a common application in any GIS system, however simple it may be. Measuring distances is important for further analysis and also to provide means of estimating travels (DeMers, 2000)

But any measurement from GIS is an approximation as vector arcs are made up of straight line segments (lines that appear as curves on the screen are stored as a collection of short line segments) and all raster entities (point/line/area) are basically one or more grid cell representation. In raster GIS, distances are measured using Euclidean or Manhattan distance. The latter is named after the city, Manhattan, which is full of dense blocks of buildings and a person has to travel by the sides of these blocks to get across the city (Fig 6.1a and b). From the figures it could be observed that a segment AC has got two lengths depending upon which method is followed for measurement. Euclidean distance is also used in case of vector GIS in measuring the distance. Here Pythagorean theorem is used to get the Euclidean distance. In vector structure, the points (for which the distance is measured) must exist in the coverage. Thus it becomes difficult to calculate/perform any distance operation for the intervening space between these points, unless some points are defined there in. For perimeter measurement, the number of cell sides (in raster) making the boundary of an area is counted and multiplied with the resolution of the grid. Similarly, for area calculation, the area of a grid-cell is multiplied with the number of cells that make the area. In this process there is always some approximation, as an entity can never be exactly made up of only whole cells. For perimeter measurement in a vector GIS the geometry is used, where the area is divided into simple geometric shapes, calculating their area using prescribed formula and then integrating them to get about the whole area (Fig. 6.1c).

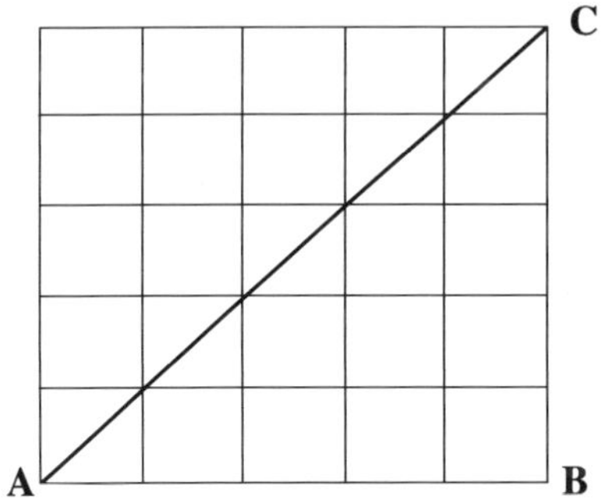

(a) Euclidean distance AC = $(AB^2+BC^2)^{1/2}$ = 7.07 units

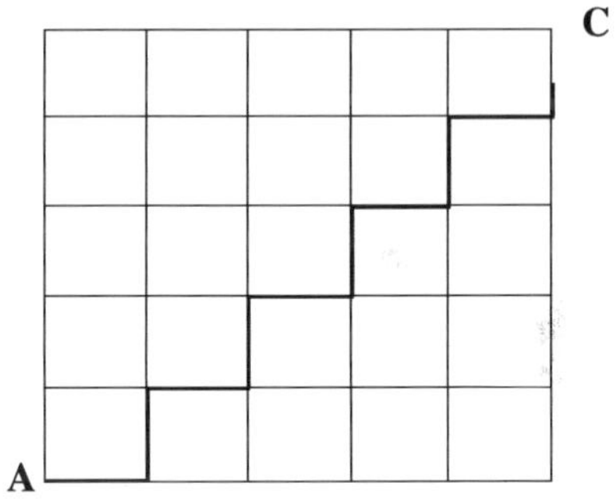

(b) Manhattan distance AC = 10 units

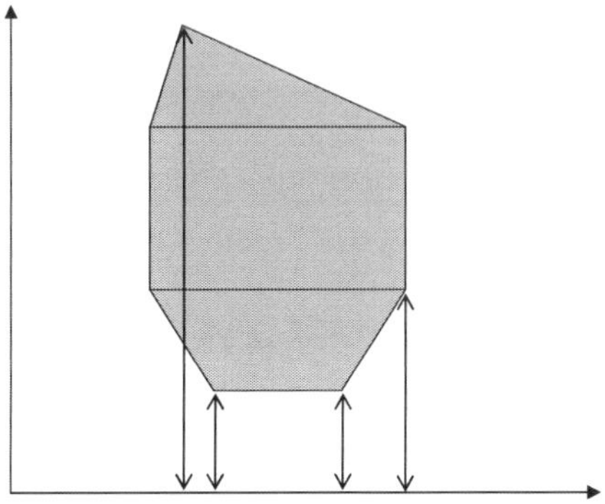

(c) Area measurements in a vector GIS

Figure 6.1: Measurement in GIS

The above is true for an isotropic surface where we consider no friction or physical barrier for the movement to be restricted. If there is restriction, we call it as non-isotropic surface and need to calculate the functional distance. Here friction refers to some type of impediment to movement or flow (i.e., terrain). Relative barrier is similar to friction but occurring for a specific area (e.g., a river, a ridge) whereas absolute barrier is an impasse that must be circumvented (e.g., a lake, a restricted area). Each of these can be assigned impedance values depending upon how difficult they are to travel through.

6.2.1 Least cost path

Least cost path can be calculated by GIS using the following steps:

1. A cost surface is created that gives the cost of passing through each cell. For example, the elevation data for each grid cell together make a topographic surface. This can be a weighted combination of several data layers (multiple criteria for a project).
2. An origin is selected.
3. All cells (grids) are set to values, sufficiently high to exceed the maximum number designated in a GIS program (DeMers, 2000). Then we move iteratively from cell to cell and the cost of moving is progressively accumulated. If the value in a cell is higher than the cost calculated, it is replaced with the new lower cost. That same cell will be revisited through other paths and in case, may be replaced by still lower cost.
4. Now a destination is selected.
5. Finally the least cost path between origin and destination is calculated.

$$\text{Accumulated distance} = 0.5 * (\text{Dist} * \text{Cost}_{x,y}) + 0.5 * (\text{Dist} * \text{Cost}_{x-1,y}) + \text{accumulated distance for cell x,y.}$$

where, Dist = 1.0 for adjacent cells and 1.414 for diagonal cells;

$\text{Cost}_{x,y}$ = cell coming from; and

$\text{Cost}_{x-1,y}$ = cell going to.

6.2.2 Shape measures

Shape is a very important concept in GIS analysis. A very steep road curve (shape of a line) may lead to accident; a sharp stream curve may invite high degree of erosion. Thus, the sinuosity (ratio of a straight-line distance to the sinuous distance) of a linear feature helps to understand the relationships between objects and their environment (Lee and Sallee, 1970).

1. Perimeter/Area Ratio: Mostly used in morphometry analyses for hydrological characterization of watershed or river basin, this ratio provides a measure of the compactness of the polygon concerned.

2. Another parameter to measure the shape is called 'Convexity Index'. Here, assuming the circle as the most compact geometric shape, any other polygon must have perimeter greater than that of a circle. In a vector GIS

   Convexity Index (CI) = kP/A

   where, k = a constant based on the size of the circle that inscribes the polygon

   P = perimeter and

   A = area of the polygon

   Convexity Index provides a measure of compactness or edginess of the polygon and gives an account of the convexity of the polygon versus that of the circle. This index may range from 1 (far from a circle's shape) to 99 (closest to a circle, but not a circle — a circle has a CI value of 100). In a raster GIS, the area is represented by the square root of the number of cells constituting the polygon. The range of values is the same i.e., 1 to 99.

   There are other measures of polygon shapes such as axial ratio (major vs. minor axis), perimeter and area of polygon singly or in combination, circularity mean side, variance mean side etc. (Davis 1986, Baker and Cai, 1992, McGarigal and Marks, 1994, DeMers 2000). All these measurements relate to the shape to the boundary configuration. There is another way to measure the boundary configuration called ***edginess*** (Berry, 1993), which uses a filter. The application is mostly centered on digital remote sensing data processing.

3. Euler Function: This is a measure of the perforation or fragmentation of an area. This function is based on the Euler's (pronounced as 'oiler') theorem, after Leonhard Euler (1707-1783), the famous Swiss mathematician. This measures numerically the degree of fragmentation or the amount of perforation of an area (i.e. whether the area is uniform or is broken up by numerous intervening inclusions). Euler number is calculated as:

   Euler number = (holes) – (fragments – 1), where 'holes' represent the number of self-contained polygon perforations in the outside polygon and 'fragments' refer to the number of polygons in the fragments region (De Mers, 2000). Let us take an example here The Figure 6.2. tells about four cases. In the first case, the surface has three fragments but no holes and the Euler number works out to be [0 – (3 – 1)] or – 2. In the second case, the polygon has three holes but no fragment; the Euler number is [3 – (0 – 1)] or 4. In case of the third, the Euler number is 2 (two fragments and three holes); and in the last case, the number becomes 5 (four fragments and eight holes). Thus for different combinations of holes and fragments, Euler number may range from – 7 (one hole, nine fragments) to + 9 (nine holes, one fragments) and in this way, effectively represent the configuration of polygons (DeMers, 2000). This may also be referred to as 'spatial integrity' of polygons.

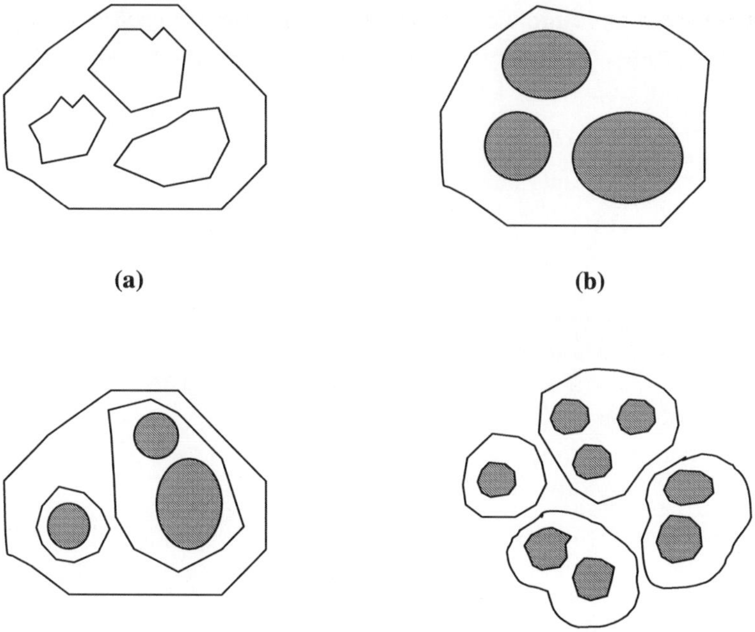

(a) (b)

(c) (d)

Figure 6.2: Spatial integrity of polygons

6.3 QUERY

One of the fundamental functions of a GIS is to retrieve selective information from its database (already stored) to be displayed in an understandable way. We may have two query types as (a) query by attribute, and (b) query by geometry.

6.3.1 QUERY BY ATTRIBUTES

We know that GIS stores spatial and attribute data in two separate files. Corresponding records in the two files (for example, a cropping system map of an area with the corresponding attribute information like name of the crops, respective areas under production, productivity of crops, etc.) are linked by an identification number (ID). The attribute database, in general, is stored in a table (relational database) mode with a unique code linked to the geometric (spatial) data points. This linkage enables a GIS to search and display any of the stored attribute data based on some spatial information, and vice versa without altering the existing data. Users can seek to retrieve the information (attributes) specific to a particular cropland (spatial location). Conversely, map features can also be retrieved on the basis of attributes e.g., we can have information of the areas with particular crop (say, rice) productivity greater than 3t ha^{-1}.

The GIS can handle even more complex queries involving rational, arithmetical and Boolean or logical expressions like +, –, =, not =,<,≡, >, ≥, and, or etc. This is accomplished by building a *query statement*. Several query statements can be combined to perform a single query. For the example cited above, one might wish to know what may be the extent of area under *kharif* rice (there may be other summer and *boro* rice) cultivated with a specific hybrid variety with productivity greater than 4t ha^{-1} and also can be spotted within 10 km from the rice-mill. Many GIS include a sophisticated function of relational database management system (RDBMS) known as Standard Query Language (SQL), to answer this type of complex queries. Using SQL, the GIS would first build and send a query statement to its DBMS, which will search the attribute database for those land parcels meeting the criteria. The DBMS then returns the ID numbers to the GIS and the GIS searches its spatial data file to find these land parcels and their coordinates, and finally highlights those parcels for their display in their same map.

The conditional statement is represented by the rational (>, <, =, ≥, ≤), arithmetic (+, –, *, /) or Boolean (AND, OR, NOT and XOR) operators. Suppose set A and set B indicates girl students in a college and students of the same college appearing in the final examination, respectively. In this case, the Boolean statement 'A AND B' represents the girl students appearing for the final examination, while all the students in the college are represented as 'A OR B' (Fig.6.3). Similarly, boys appearing in the final examination are designated as 'B NOT A' and number of boy students appearing and number of girl students not appearing in the examination as 'A XOR B'. This type of queries may also be referred to as *set queries* as they involve two or more sets.

Logical operators can as well be employed to narrow the search or even to specify which attributes to retrieve. The resulting query set could be used to generate reports, graphic displays, and new map files. Queries are also useful to check the integrity of the data and the results at any point of analysis.

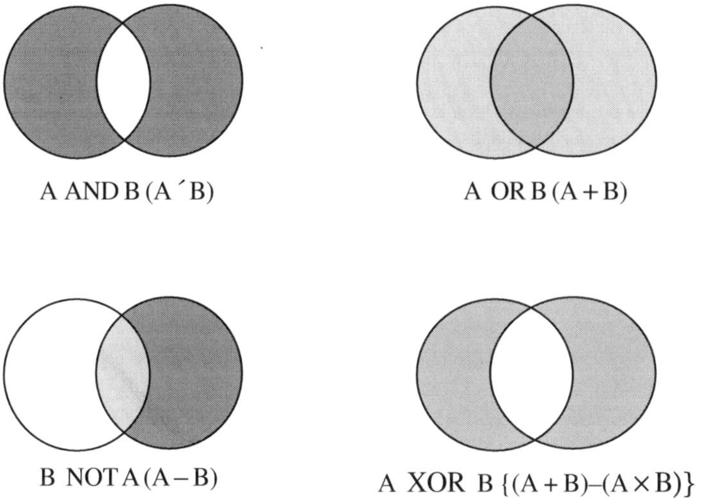

Figure 6.3: Boolean operators

6.3.2 Spatial queries

This type of queries are performed through a query set based on the spatial relationship of map features. Spatial operators in the queries define the spatial relationships that exist between map features. Spatial operators (e.g., ***overlaps**, **entirely contains/entirely contained by, contains/ contained by**, **terminates in/terminus of**, **passes through/passed through by**, **on boundary of**, **touches/meets**, **spatially equal etc.***) can be combined to answer complex spatial queries. For example, nation's future flood management program might require a database of land parcels and classify them into flood-prone and non-flood-prone areas. GIS can be used here to compare past flood affected areas to the land parcel map. First of all, the user would make query for all parcels that falls within the flood plain area for most of the years (using the 'entirely contains' operator). These addresses would be classified as *flood-prone*. The system would then be used to find parcels crossed by the flood plain boundary (using the 'passes-through' operator). Address in this category may be classified as ***intermediate*** and should undergo further investigation to determine the exactness of the area falling under flood-prone category. The remaining parcels would be entirely outside the flood plain and can be put into the *non-flood-prone* category.

6.4 Attribute Based Operation

The GIS finally provides output in the form of map or graphical text (Chrisman, 1997). For example, a hydrologist may want to know the runoff generation potential of a watershed to compute the water balance of that region; an ecologist might be interested to quantify the biodiversity in a region; an economist may want to know the demographic composition of a village. These attribute information in the final form are the results of the interaction of spatial and attribute data contained in a GIS. In this process total information content may increase / improve or some information may be lost at the cost of generalization. Based on the information content at the final stage (output) compared to the parent data, these operations can be any of the two as follows:

(a) analysis where information content is reduced
(b) analysis where information content is increased

6.4.1 Reduced Information Content

a. Generalization or grouping Attributes can be generalized by merging similar sub-classes e.g. a detailed land use / land cover classification that includes many types of land classes and can be grouped together in a few classes. In a study with the objective of generating runoff potential of an area, the land use/land cover classes were grouped into three (poor, moderate and good) to apply curve number method in a GIS (Chakraborty et al., 2005). This was obtained through merging of polygons in the spatial domain by dissolving polygon boundaries.

b. Selection It is often desirable to select a set of entities based on certain criteria (a combination of their attribute data). For example, to know the erosion status of a region we need to select soil types that contain more than a certain amount of silt, are poorly drained and on >15% slope. SQL is the standard data base query language for this purpose (i.e., select soil units which contains >30% silt, drainage status = 'poor' and slope is <10%).

Boolean operators (AND, OR, XOR, NOT) are used for complex selection from an entity with multiple attributes, e.g., if A is a set of soil units having clay loam texture, B with pH > 7.0, and C are units that are poorly drained, then we can get many combinations like,

X_1 = A AND B

X_2 = A OR B

X_3 = A XOR B

X_4 = A AND (NOT B)

X_5 = (A AND B) OR C

X_6 = A OR (B AND C), where each of these will fulfill some or other criteria.

c. Classification / Reclassification Classification is a procedure that creates ordinal or nominal categories from ratio or interval data (Chrisman, 1997). This is an age-old and common phenomenon in our day-to-day life. It is basically to arrange things in a framework for proper understanding of their behaviour and their changes in a complex dynamic world. We classify objects in a way we wish to view them and apply in future, and thus involve many ways and means to do so. This may as well be influenced by the scale or resolution as by the methods/ instruments having employed in the classification. The criteria set for building the categories seem to be simple for classifying the land surface feature (e.g., cropland, wasteland, water body, forests etc.) but the practical aspects of the task is far from straightforward (DeMers, 2000). This probably explains the vast number of possible classification and reclassification methodologies — all depending on the needs and objectives of the user.

Land classification types as available today vary from classification for different land surface features separately (vegetation, soils, geology) or for prospective use of these features (agriculture, urbanization). Since the advent of satellite remote sensing, a very dynamic and efficient mode of information generation has come into existence. With GIS offering an otherwise impossible task of analyzing the huge data as required by the user, the importance of classification in decision-making could be amply understood. GIS made it possible to use the same raw data to produce specific output. For example, the soil and land use maps developed by National Bureau of Soil Survey and Land Use Planning (NBSS & LUP) can be effectively combined in a GIS to generate land suitability classification; or the UNSCO classification system (1973) for the world vegetation can be manipulated to functional ecosystem mapping of a nation. This is probably the essence and the most important point about classification using GIS (DeMers, 2000).

Reclassification involves reassigning new thematic values (or codes) to the spatial features, producing new images. In a raster GIS, it is simply changing the legends to represent the new coverage. And a word of caution—while changing the attribute codes one must be careful enough of not assigning the existing values again.

In vector GIS, the same task requires change in both entities and attributes (e.g., line dissolve). The common boundaries between two or more polygons are dissolved followed by merging of selected polygons. The new values can subsequently be assigned to represent a new classification. This type of 'reducing' or 'aggregating' data (i.e. simplifying the coarser information) is very much useful to compare between two coverage to analyze a decision. For example, erosion potential of different land use/ land cover types varies widely. Thus, an area with several land use types can be reclassified based on the erosion potential of individual land use type. The new image thereby indicates erosion potential classes (high, moderate or low) rather than general land use / land cover classes. Let us examine the following cases to understand reclassification more clearly. All these examples stand true for nominal data, but it is altogether the same for other type of data also (interval / ratio ordinal) (Fig. 6.4).

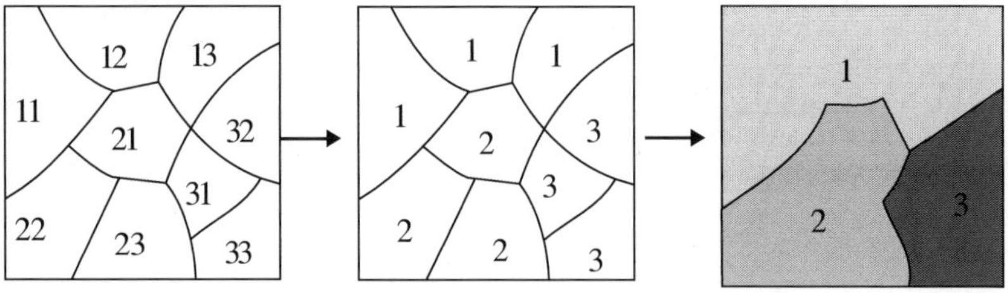

(a) Generalization with reduced classes

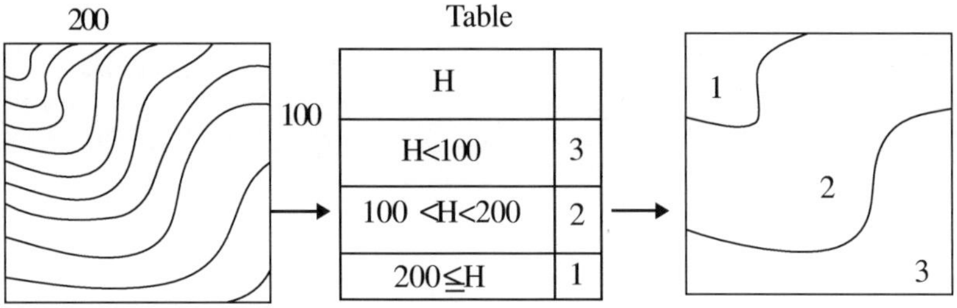

(b) Ranking with classification table

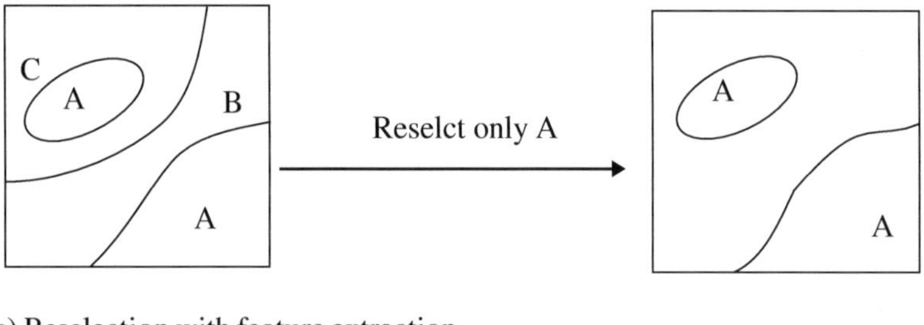

(c) Reselection with feature extraction

Figure 6.4: Reclassification

(i) Generalization

As discussed, generalization refers to the reassignment of existing data into smaller number of classes. This results in a reduction of the level of details. For example (Fig. 6.4a), areas under crops, orchard and seasonal fallow (11, 12 and 13, respectively) are grouped together as agricultural land (reclassified as 1). Similarly, residential plots, commercial plots and lake (21, 22 and 23, respectively) as non-agricultural mixed land and open, semi-open and close forest areas (31,32,33, respectively) can be grouped as forest land. Hence, the reclassified new image shall consist only three land use / land cover types (as compared to nine in the old image).

(ii) Ranking

Ranking refers to valuation of attributes based on an evaluation model or a specified table. Suppose, we need to identify and characterize contours at 100 m interval, in place of 10 m interval as found in the original document. This can well be done using reclassification (Fig. 6.4b).

(iii) Reselection

Some times we need to select features of our interest to be kept and unselected features are removed. To retain only the agricultural land (A) in the final output, we go for reselection and remove other classes (Fig. 6.4c).

The reclassification as explained above is on the basis of locational information, which is only one of the four basic methods (the other methods are based on position, size and shape information) (DeMers, 2000).

6.4.2 INCREASED INFORMATION CONTENT

The previous section discussed the reclassification as a procedure where the total information content is apparently reduced; we come down to fewer categories than before. Conversely there are analyses, which increase the information content in the output. The following are some of the examples:

a. Rank: Ranking increases information content when categorical data is converted (elevated) to ordinal ranking (i.e., forest cover transformed into habitat quality).

b. Evaluating: This is the conversion of categorical data to interval or ratio data using the categories as indirect measures of other phenomena (i.e., forest cover data is evaluated for species density estimation).

c. Combining Pairs of Input Values: This involves operations for interval or higher levels e.g., sum and difference (i.e., total population is summation of each group of people) or operations on ratio measurements like proportions (i.e., men and women to total population), density (i.e. number of people per unit area) and rate (increase in population over last 10 years).

6.5 Neighborhood Analysis

In reclassification, we change the values of a pixel or the attributes contained therein. But in nature every point is a function of its neighboring points. Hence, a locational change should be done keeping in view of its position within its surroundings. This can be achieved by using neighborhood functions. It involves the alteration of values of a pixel on the basis of the adjacent pixels. In neighborhood analysis we usually define a target area and its neighboring areas resulting in new coverage. This is therefore called 'targeted' analysis of neighborhood as against 'total' analysis when entire area is taken into account (DeMers, 2000). For targeted analysis, the neighbors are the immediate cells adjacent to the target cell. It is static in nature as the whole analysis is performed at one go.

a. Roving Windows or Filters: *Filtering* is a kind of neighborhood analysis, which is very common in digital image processing. The filter is defined as a group of cells around a target cell (Heywood et al., 2000). The user usually sets the number of cells. Common filter shapes are square and circular and the dimensions of the filter determine the neighboring pixels used in the filtering process. Filter may be ***high-pass*** (used to measure edginess of polygon features) or ***low-pass*** (used to remove unusual high / low values) or ***directional*** (used to measure orientation of pixels). For example, one has to find out small ridges in a watershed to study the prospects of rain water harvesting in details. A high-pass filter with a high value at the centre pixel and very less values in surrounding pixels (in a 2×2 or 3×3 matrix of grid) may be used to enhance the higher values even higher and suppress the lower values even lower. This results in a coverage that depicts the ridges (even small) more clearly. This process is especially useful to work at rapid changes in elevations. Similarly for smoothing out an elevation model that might have spurious elevations, a low-filter is the best option. Directional filters are used to highlight the ridges (for example) which have desirable orientation (north-south or east-west).

b. Trend Surface Analysis: This refers to analysis of the change in surface direction and slope, which gives an idea of the trend of changes as occurring over an area of interest. Slope (S) and aspect (A) are the two basic components in this analysis. Slope is a measure in the change in elevation over the cell (measured as an angle in degrees or as a percentage) whereas aspect is the direction of the slope (usually expressed in degrees from north).

Mathematically,

$$\tan S = [(-dz/dx)^2 + (dz/dy)^2]^{0.5}$$

$$\tan x = (dz/dy) / (dz/dx)$$

If Aspect = A

then A = 90 (–) x if (– dz/dx, – dz/dy);

A = 90 (+) x if (– dz/dx, + dz/dy);

A = 270 (+) x if (+ dz/dx, – dz/dy); and

A = 270 (–) x if (+ dz/dx, + dz/dy);

where

$$[dz/dx]_{i,j} = [(Z_{i+1,\ j+1} + 2Z_{i+1,\ j} + Z_{i+1,\ j-1}) - (Z_{i-1,\ j+1} + 2Z_{i-1,\ j} + Z_{i-1,\ j-1})]\ /\ 8dx$$

$$[dz/dy]_{i,j} = [(Z_{i+1,\ j+1} + 2Z_{i,\ j+1} + Z_{i-1,\ j+1}) - (Z_{i+1,\ j-1} + 2Z_{i,j-1} + Z_{i-1,\ j-1})]\ /\ 8dy$$

(dx = distance between cell centroids in the x direction and dy = distance between cell centroids in the y direction) (Fig. 6.5).

$Z_{i-1,\ j+1}$	$Z_{i,\ j+1}$	$Z_{i+1,\ j+1}$
$Z_{i-1,\ j}$	$Z_{i,\ j}$	$Z_{i+1,\ j}$
$Z_{i-1,\ j-1}$	$Z_{i,\ j-1}$	$Z_{i+1,\ j-1}$

85	90	100
80	85	90
70	80	90

Figure 6.5: Calculation of gradient and aspect

Thus, for the elevation values for the pixel as given in Figure 6.4b,

$$[dz/dx]_{i,j} = [(100 + 2*95 + 90) - (85 + 2*80 + 70)]\ /\ 8 * 10 = 0.8125$$

$$[dz/dy]_{i,j} = [(100 + 2*90 + 85) - (90 + 2*80 + 70)]\ /\ 8 * 10 = 0.5625$$

Gradient (G) will be

$$\tan G = (0.81252 + 0.56252)**0.5 = 0.988 \sim 44.65 \text{ degrees}$$

and Aspect (A)

$$x = \text{Arctan}\ (0.5625/0.8125) = 34.7$$

$$\text{Aspect (A)} = 270 - 34.7 = 235.3 \text{ degrees}$$

After calculation of slope and aspect, other complex parameters like rate of change in slope, convexity or curvature of slopes are generated for a complete terrain / surface analysis.

c. Buffering: In GIS, a *buffer* refers to a polygon created at a distance from a point, a line or an area. Buffering is a typical neighborhood function that is used for identifying areas surrounding geographic features. This process involves generating a buffer around existing geographic features (point, line or area) and then identifying or selecting entities based on whether they fall inside the boundary of the buffer. Thus it is merely an extension of the specified distance measurement technique as discussed earlier (Section 6.2). Nevertheless, with all its importance, buffering process should not be equated with simple measurement in GIS and most of the GIS today include this function as a separate module. Figure 6.6 shows buffer zones (distance specified as *r*) around a point (a circle with radius *r*), a line (a rectangle of width *r* having rounded ends) and an area (irregular shaped path with width *r*). Buffer zone may be a temporary polygon for specific query or may permanently be included in the database, but it is always stored as a separate coverage. Thus any number of operations can be performed with other data layers.

Buffers are often used in situations where protection of a resource is needed (buffering seems to prevent timber cutting), in finding the existence of certain entities with respect to other entities (parking sites within 0.5 km of office areas) or in demarcating a zone (100-year flood zone). An example is discussed as follows.

A study has been carried out to monitor the vegetation status along drainage buffer of an arid watershed in Rajasthan, India (Chakraborty et al., 2001). As the drains in the study areas are ephemeral in nature, one would expect moisture for maximum period of the year along the drains, than anywhere else, which in turn, would result in better vegetation status. The idea behind this was to confirm any change occurred due to drainage line treatment like check dam, loose boulder dams, grassy waterways, *nalla* bunds etc. within the preview of the watershed development project guidelines covering the study area. It was assessed that any activities mentioned above will arrest sediment load and will facilitate better infiltration opportunity time, which could be reflected through vegetation along the recharge zone. Hence, a buffer region of 50 m on either side of the drains was taken to evaluate the changes in the vegetation vigor (based on NDVI values derived from satellite remote sensing data) along the drainage lines.

Some of other typical examples of buffering analysis may be:

- Determination of the population size that resides within 50 km from the sea shore;
- Computation of the number of houses that is within 100 km from the national highway;
- Monitoring the vegetation status along a drainage network;
- Determining of the number of hotels within 10 km of a main road.

Raster vs. Vector Buffering: Conceptually buffering is very simple to understand but very complex to operate. Creating buffer in raster is rather simple as we need to consider only the number of pixels (grid cells) away from the features. Raster buffering provides a unique distance for each cell. This approach can better be called as *proximity analysis* (see section 6.6.1) than a buffering. In contrast, while doing buffering operation in vector data structures, one has to remember that the topological information must be explicitly coded for each polygon produced as buffer. This needs a careful lookout of the software options available with the vector GIS.

There are four basic types of buffer (DeMers, 2000) as *arbitrary* buffer (where the boundary of buffer is made through some guess or arbitrary knowledge without concrete scientific support base), *causative* buffer (some priori knowledge is available), *measurable* buffer (based on definable and measurable values) and *mandated* buffer (based on some legal or mandated measures). The buffer need not be of same distance all through the path round an entity, and many times the width of the path varies depending on the barriers or obstacle encountered.

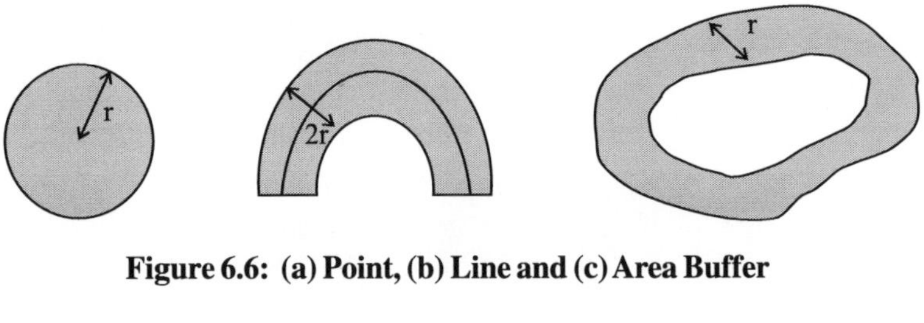

Figure 6.6: (a) Point, (b) Line and (c) Area Buffer

6.6 Connectivity Analysis

It deals with the operations depending on the connectivity between points, lines and polygon in terms of distance, area, travel time, optimum paths etc. Here the entities are directly linked in the database. These analyses basically include proximity analysis, network analysis, etc.

6.6.1 Proximity analysis

Proximity analysis is the way of analyzing locations of features by measuring the distance between them and other features (points, lines and boundaries of polygons) in the area. One of the most popular proximity analyses is based on the process of *buffering* (see Buffering). The distance between two points may be measured as a straight line or by following a network path, such as a street network. The sinuous path is often the ultimate choice in the real world situation. Thus the practical distance between two locations follows the path through which one can travel and need not necessarily be the straight-line distance between them. For example, an industrialist is interested in building a manufacturing plant, where he considers the distance from the nearest railway station and the airport. Through GIS he will try to retrieve the information by simply clicking on the point locations representing the site and the railway station or airport. Once the distances are determined, other pertinent information such as water and sewer availability, price per unit land, and availability of labor can be analyzed from the existing database.

Proximity analysis can also be based on time. For example, proximity analysis based on access time or travel time will concentrate equidistance time zones around a certain point (A). The resulting coverage indicates the time required to reach 'A' from another point at some distance. The new coverage is thus like a distance surface which grows outward from the feature of interest (Berry, 1987; Tomlin, 1990; Heywood et al., 2000). This coverage can be subsequently reclassed to create a causative or measurable buffer zone.

6.6.2 Network analysis

A network is a set of interconnected lines. We have already learnt how to construct the network data structure. Here we will consider the applications of the network data structure using a GIS. The network analysis is a special type of connectivity-of-line analysis involving a set of interconnected lines (roads, streams, railway tracks, sewage lines). Network analysis can be used to address at least four types of queries.

1. Address geocoding: This is a process of taking locational addresses and estimating the locations in GIS. The process requires a table of addresses (mostly the addresses of buildings or houses) and the theme (e.g. street theme) containing the attributes that can be used to match to the table of addresses. Here, each street must contain the beginning and end addresses on the left and right sides of the street (Fig. 6.7).

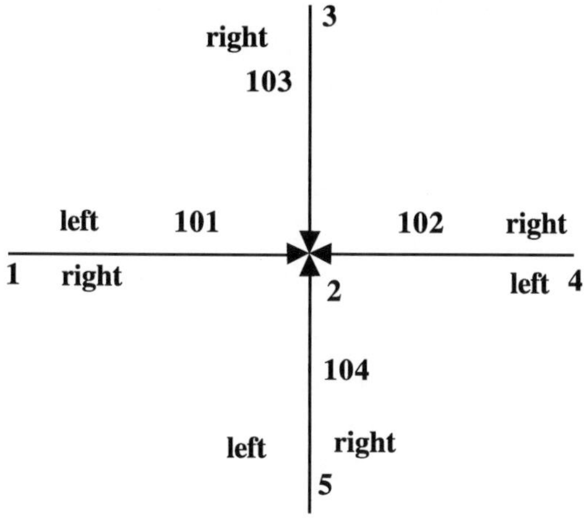

Arc#	F_node	T_node	Length (m)	L_from	L_to	R_from	R_to	Prefix	Name	Type	Suffix
101	1	2	961	5	112	10	135	West	B.D.	Marg	
102	4	2	505	32	155	120	24	East	K.G	Marg	
103	3	2	440	3	29	31	49		Sardar	Street	North
104	5	2	520	39	71	62	40		Library	Street	South

Figure 6.7: An example of address geocoding

Suppose we want to reach to a house with address 40 East K.G. Marg. GIS first determines which arc contains the prefix/name/type/suffix. Here it is arc 102. The number 2906 address can be estimated using linear interpolation. As it is an even number, this must be (say) on the right side of the arc 102. The address ranges from 120 to 24 (range 96) along the right side of the arc. Length of the arc is 505m. Thus the address 40 can be geocoded as:

[Max. address - queried address] / range] *arc length = [(120-40) / 96] *505 = 420 m along the arc.

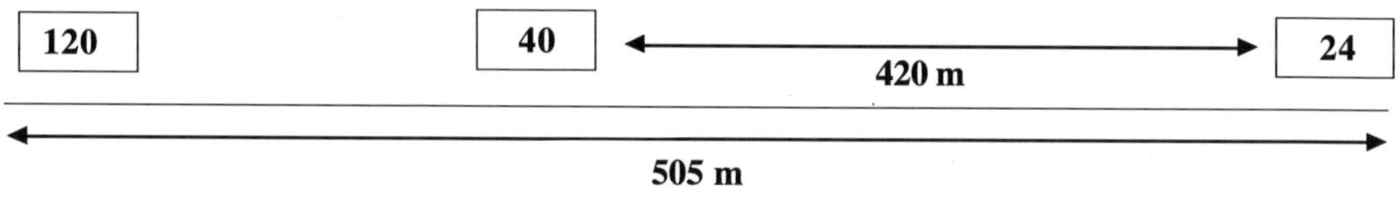

2. Optimal routing: This is another type of network analysis wherein the best route (path) to get from one location to another location can be delineated. A route is a collection of arc segments (or sections) that have attributes. A collection of routes is called a route system. Optimal routing identifies the most efficient routes or paths using specified decision rules. The rules are likely based on minimum time or distance, maximum correlation occurrence or capacity and so on. Instead of testing all possible pathways, a path finding algorithm e.g., Dijkstra algorithm (named after E.W. Dijkstra) can be used. The *cost* in a network analysis is frequently the shortest distance or travel time to visit a location or a set of locations in a network. To get a clear picture, let us take an example where we have to find out the least time to travel from point 'X' to point 'Y' (Fig. 6.8). The values (75, 50. 10...) indicate the time in minute to travel.

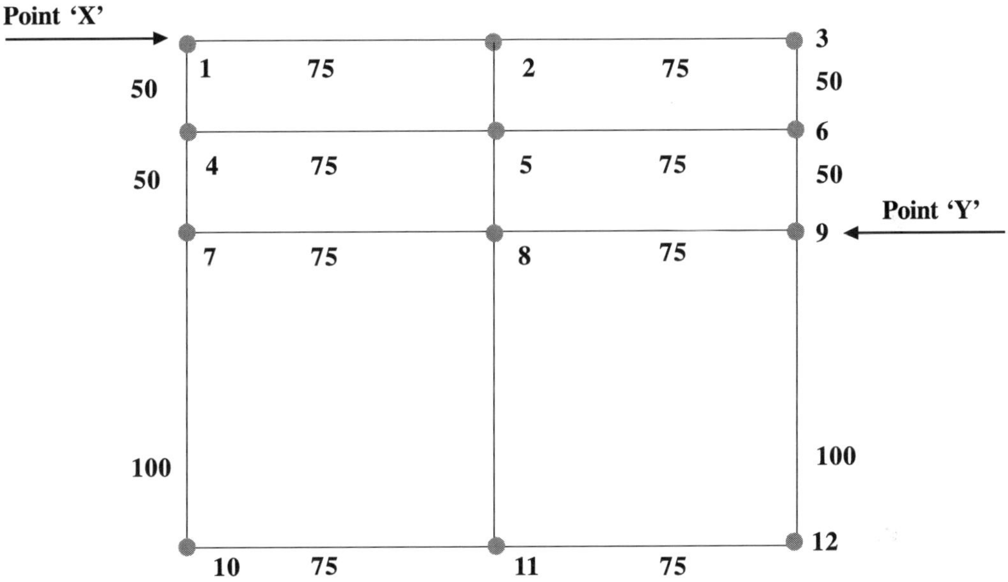

Figure 6.8: Optimal routing analysis

Step1. We build two tables–one for the nodes already been processed and another for adjacent nodes to process. Staring from point 'X' (i.e., node 1):

Processed nodes		
Node	Cumulative cost	Previous Node
1	0	

Adjacent nodes		
Node	Cumulative cost	Previous Node
2	75	1
4	50*	1

* the least cumulative cost of travel

Step 2. We pick the adjacent node with least cumulative cost (i.e., node 4) and add to the processed nodes list. Then adding the nodes adjacent to the last processed node to the adjacent nodes list......nodes 5 and 7, we get:

Processed nodes		
Node	Cumulative cost	Previous Node
1	0	None
4	50	1

Adjacent nodes		
Node	Cumulative cost	Previous Node
2	75	1
4	50*	1

Step 3. We pick node 2 (with least cumulative cost) and add to the processed nodes list. Then we add the nodes adjacent to the last processed node to the adjacent nodes list. Thus we get:

Processed nodes		
Node	Cumulative cost	Previous Node
1	0	None
4	50	1

Adjacent nodes		
Node	Cumulative cost	Previous Node
2	75*	1
5	125	4
7	100	4

Step 4. We pick node 5 (least cumulative cost) and add to the processed nodes list. After adding the nodes adjacent to the last processed node to the adjacent nodes list we get:

Processed nodes		
Node	Cumulative cost	Previous Node
1	0	None
4	50	1
2	75	1
5	85	2

Adjacent nodes		
Node	Cumulative cost	Previous Node
7	100	4
3	150	2
6	160	5
8	95*	5

By doing the subsequent steps like this, finally we get:

Processed nodes		
Node	Cumulative cost	Previous Node
1	0	None
4	50	1
2	75	1
5	85	2
8	95	5
7	100	4
3	150	2
6	160	5
11	145	8
9	170	8

Processed nodes		
Node	Cumulative cost	Previous Node
10	200*	7
10	270	11
12	220	11
12	270	9

Step (n-1). We can now pick node 10. All the nodes adjacent to node 10 (nodes 7,11) have already been processed. So we add node 12 as the final entry to the processed nodes table:

Processed nodes		
Node	Cumulative cost	Previous Node
1	0	None
4	50	1
2	75	1
5	85	2
8	95	5
7	100	4
3	150	2
6	160	5
11	145	8
9	170	8
10	200	7
12	270	11

Thus the quickest path from start point 'X' (node 1) to destination point 'Y' (node 9) is 9<8<5<2<1 and the optimal route will take 170 minutes.

3. Finding closest facilities: This is a special type of routing problem where we need to find out the closest points (facilities) to a given location (event). For example, to find out which of the two fire stations would have the least response time to report fire event or the best ambulance station to respond to the accident report etc. Here, the location of facilities must be fixed and exist as a GIS point theme. The event locations can be entered into the network by several ways (through address geocoding or through creation of an even point theme etc.).

4. Resource allocation: It is typical application of network where the allocation of resources (goods, people or services) takes place from supply centers to the points of demand in a network system; for example, a well or hand pump to supply water to the field through irrigation network. Network analysis is used here to optimize the allocation of resources. Typically resources are allotted across the network till the resources are exhausted or demand is satisfied.

For effective network analysis, the topology must be very accurate. The correct geographical representation is not always necessary, as the real world scenario is too complex to be depicted; but the topological representations should mimic exactly the real world situation (Reddy, 2001).

6.7 Overlay

One of the key functions of GIS-map overlay refers to the process of stacking digital representations of various spatial data on top of each other so that each position in the area covered can be analyzed in terms of these data (Burrough 1986). Overlay integrates multiple types of spatial data from two or more sources and creates a new data layer (output) in a way similar to overlaying of tracings of two paper maps on a light table. McHarg (1969) introduced this concept through his work on landscape architecture, though his overlaying process was manual. Overlay function has wide applications ranging from visual comparison to integration of complex operation. As with other operations, there are differences between raster and vector overlay, the first one being relatively easy, efficient, less time consuming and operationally less expensive.

6.7.1 Raster overlay

Raster overlay with more than two data layers is easier as compared with overlay of vector data, because it does not include any topological operation but only pixel-by-pixel operations. But instead of being smooth, object boundaries will be uneven due to the use of cells. Though accuracy can be improved by increasing the resolution (i.e. using 10 m cells in place of 50 m), but data sets will become too large to process.

The idea of map algebra (Mapematics; Berry, 1993) is applied in overlaying raster data where input data layers may be added, subtracted, multiplied or divided to get output data layer (Heywood et al., 2000). Other mathematical functions can also be easily performed. To understand raster overlay concept, let us consider a watershed area with three data layers as input—location of open wells as point data, roads as line data and land use / land cover as polygon data. The well locations are given a value of '1' and roads are coded as '2' in the respective data layer. In the land use map, cells with code '3' refer to agricultural land, '4' to settlements and '5' to non-agricultural lands including water bodies (Fig 6.9). All data layers contain some cells with '0' value that represent features out of interest. If we are interested to find out which agricultural land contains well, we need to perform point-in-polygon overlay with *add* operation. The resulting map will contain cells with values '0' (neither agricultural land nor well), '1' (well, but not in agricultural land), '3' (agricultural land with no well) and '4' (agricultural land with well within).

In another case, to find out roads passing through agricultural land, a line-in-polygon overlay with *add* operation can be performed. The output map thus contains cells with values '0' (neither road, nor agricultural land), '2' (only road segments passing through other areas), '3' (agricultural land, no road) and '5' (road passing through agricultural land).

For polygon-in-polygon overlay, let us take another polygon map of soil texture (sandy loam an '10', loam as '11' and sandy clay loam as '12'). If this map is overlaid over land use map, we get the distribution of soil texture within the agricultural land area. This can even help to decide and suggest alternate land use pattern through subsequent analyses. Generally there are two methods of raster-based overlay.

0	0	0	0	0
0	0	0	1	0
0	1	0	0	0
0	0	0	1	0
0	0	0	0	0

Location of wells

0	0	0	2	0
0	0	2	0	0
0	0	2	0	0
0	2	0	0	0
0	2	0	0	0

Road map

3	3	5	0	5
3	3	3	5	5
3	3	5	5	0
5	5	5	5	4
5	3	3	4	4

Land use map

3	3	0	0	0
3	3	3	1	0
3	4	0	0	0
0	0	0	1	0
0	3	3	0	0

Point-in-polygon overlay
(wells in agricultural land)

3	3	0	2	0
3	3	5	0	0
3	3	2	0	0
0	2	0	0	0
0	2	3	0	0

Line-in-polygon overlay
(Roads in agricultural land)

10	10	11	12	12
11	10	11	12	12
11	11	11	11	12
11	11	11	12	12
11	12	12	12	12

Soil texture map

3	3	5	0	5
3	3	3	5	5
3	3	5	5	0
5	5	5	5	4
5	3	3	4	4

Land use map

10	10			
11	10	11		
11	11			
	12	12		

Soil texture in agricultural land

Polygon-in-polygon overlay

Figure 6.9: Raster overlay analysis

a. Weighting Point Method Basically two layers with the values of P_1 and P_2 respectively are overlaid with the weight of w_1 and w_2 ($w_1 + w_2 = 1.0$), respectively, as follows.

$$P = w_1 P_1 + w_2 P_2$$

The weighting point method is only available when the attributes have numerical values, which can be operated arithmetically.

b. Ranking Method The attributes of the two layers are categorized into five ranks as excellent (5), better (4), good (3), poor (2), and bad (1) before a specific purpose of overlay. Then the two different layers of A and B are overlaid by following one of the three ranking tables as shown in Table 6.1.

Table 6.1 a. Minimum ranking, b. multiplicative ranking, and c. selective ranking

a.

B \ A	5	4	3	2	1
5	5	4	3	2	1
4	4	4	3	2	1
3	3	3	3	2	1
2	2	2	2	2	1
1	1	1	1	1	1

b.

B \ A	5	4	3	2	1
5	5	5	4	3	2
4	5	4	3	3	2
3	4	3	3	2	1
2	2	3	2	2	1
1	1	2	1	1	1

c.

B \ A	5	4	3	2	1
5	25	20	15	10	5
4	20	16	12	8	4
3	15	12	9	6	3
2	10	8	6	4	2
1	5	4	3	2	1

Minimum Ranking: Lower rank is taken as the new rank of the overlaid pixel as the safety rule.

Multiplication Ranking: Two ranks are multiplied because of more influential effect rather than additional effect.

Selective Ranking: Experts can set up combined ranks depending on professional experience.

For practical purposes, users should build a model of overlay with many layers and the hierarchical structure as shown in Figure 6.10 (Chakraborty et al, 2004).

6.7.2 Vector Overlay

Overlay of vector data is little bit complicated because it must update the topological tables of spatial relationships between points, lines and polygons. Overlay of vector data results in the creation of new line and area objects with additional intersections or nodes that need topological overlay.

Broadly, vector overlay operations can be classified via two methods:

1. Through the elements contained in the layers to be overlaid (i.e. whether the layers contain point, line or polygon elements).

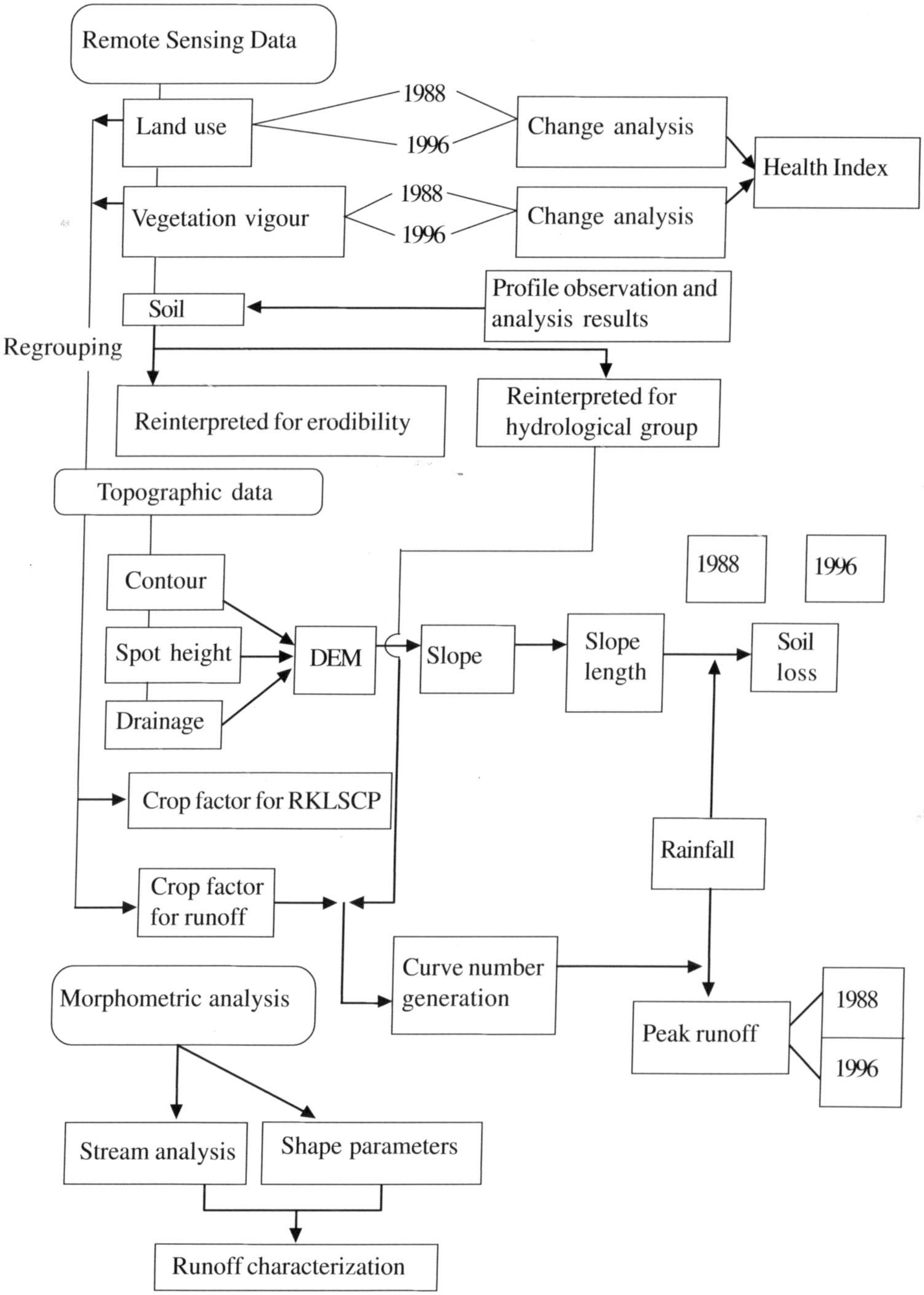

Figure 6.10: A modal of overlay for watershed hydrology study

2. By operation type (for example, the user wants to generate a layer comprising the Union, Intersection, or some other Boolean operation of the two input layers).

When classifying the vector overlay operation via method one, the element types contained in each layer are considered. The following table (Table 6.2) identifies which overlay option exists for each possible combination of element types contained in the two input layers.

Table 6.2 Types of overlay operations

Input layer element types	Points	Lines	Polygons
Points	Points Coincide	Point on Line	Point in Polygon
Lines	Point in Line	Line Intersection	Line in Polygon
Polygons	Point in Polygon	Line in Polygon	Polygon Overlay

1. Point in polygon overlay: Points are overlaid on polygon map as shown in Fig. 6.11a. Topology of point in polygon is the "is contained in" relationship. Point topology is a new attribute of polygon for each point.

2. Line on polygon overlay: Lines are overlaid on polygon map with broken line objects as shown in Fig. 6.11b. Topology of line on polygon is a "is contained in" relationship. Line topology is the attribute of old line ID and containing area ID.

3. Polygon on polygon overlay: Two layers of area objectives are overlaid resulting in new polygons and intersections as shown in Fig. 6.11c. The numbers of new polygons are usually larger than those of the original polygons. Polygon topology is a list of original polygon IDs.

Complex databases such as GIS classify vector overlay operations via method two, where the particular overlay operation a user wishes to perform, defines which element types may be contained in the two input layers.

6.8 Coverage Rebuilding

Coverage rebuilding is a boundary operation to create new coverage that are identified and selected by users. Boundary operations include the following six commands (Fig. 6.12):

1. *Clip:* to identify and preserve features within the boundary-of-interest specified by users. It is also called a *cookie cutter.*
2. *Erase:* to erase features inside the boundary while preserving features outside the boundary.
3. *Update:* to replace features within the boundary by cutting out the current polygons and pasting in the updated polygons.

4. *Split*: to create new coverage by clipping geographic features with divided borders.
5. *Append:* to merge the same feature classes of points and lines from the adjacent coverage.
6. *Map join*: to join the adjacent polygon features into a single coverage and to rebuild the topology. It is popularly called mosaicing.

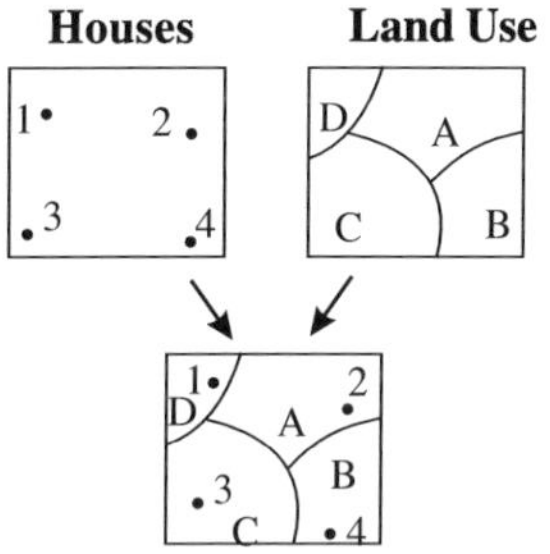

(a) Point in Polygen Overlay

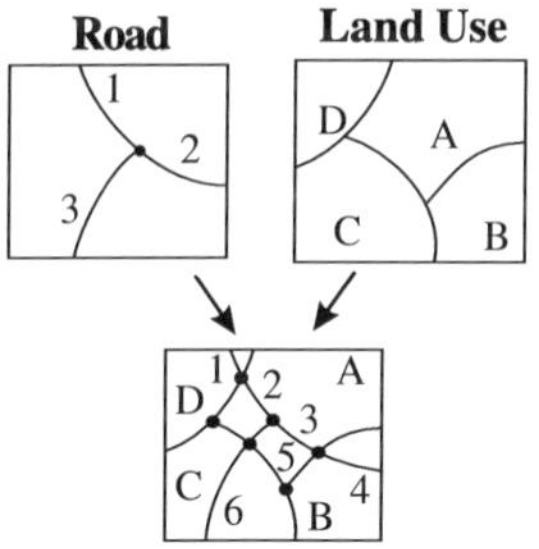

(b) Line on Polygen Overlay

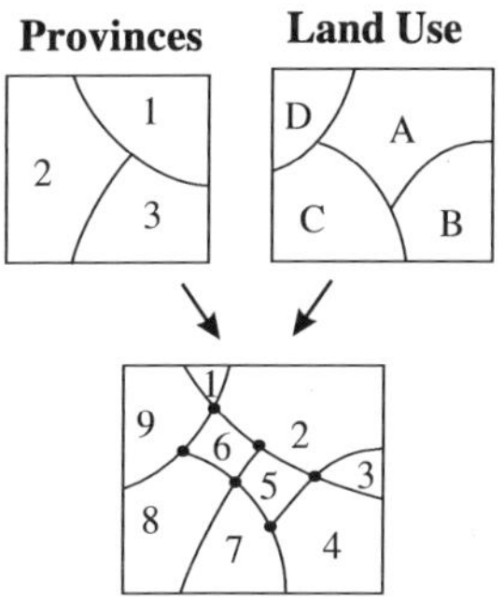

(c) Polygen on Polygen Overlay

Point Topology

ID	Owner	Land use
1	Bose	D
2	Sharma	A
3	Singh	C
4	Ahmed	B

Line Topology

ID	Road	Land use
1	1	D
2	1	A
3	2	A
4	2	B
5	3	A
6	3	C

Polygon Topology

ID	Provinces	Land use
1	1	D
2	1	A
3	1	B
4	3	B
5	3	A
6	2	A
7	3	C
8	2	C
9	2	D

Figure 6.11: Overlay of vector data

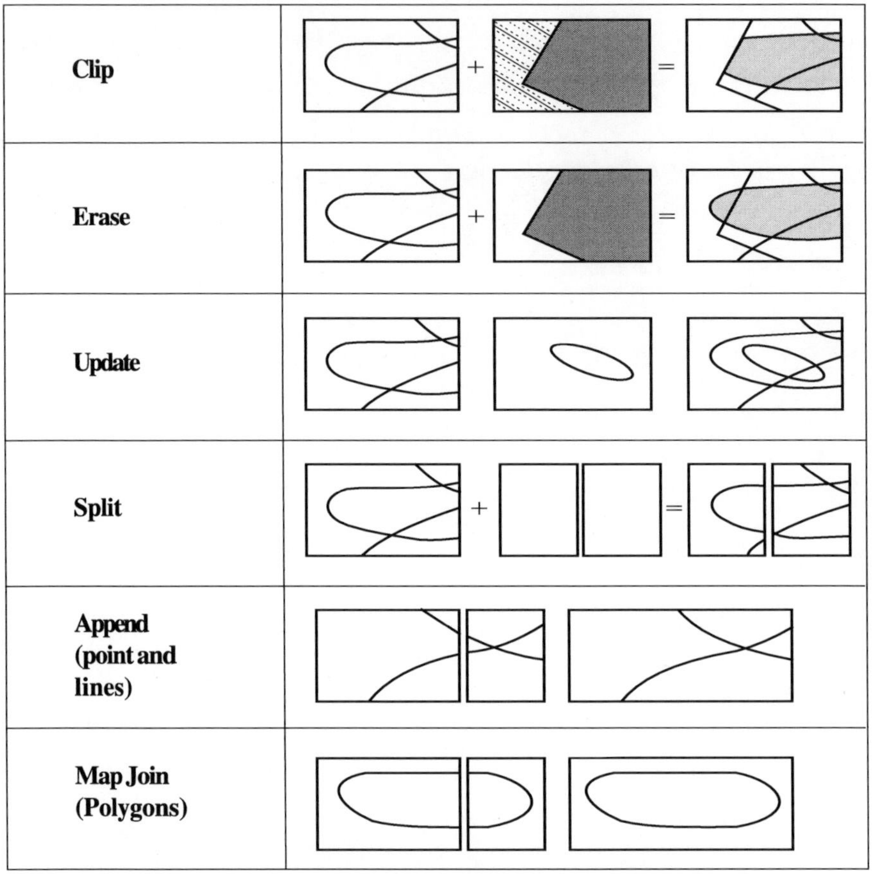

Figure 6.12: Concept of coverage building

CHAPTER 7

MODELLING SURFACE

"The new source of power is not money in the hands of a few but information in the hands of many."

John Naisbitt, Megatrends

7.1 THE ROLE OF INTERPOLATION

The main strength of GIS is the analysis of compound and large volume of spatial and attributive data. These data are collected through samplings. The values of the collected type of data can be expanded to the site points where samples are not available, using *interpolation* methods. The interpolation can be performed either in the phase of data production or when the data are used for spatial analysis; the latter is the case of a GIS function. Spatial interpolation is based on the notion that points close together in space are likely to have most similar values than points far apart (Tobler's Law of Geography). This is also a positive spatial autocorrelation between two points in space. For example, two points which are a few meters apart usually have the same altitude than those having few kilometer distances between them; or two soil profiles in the same plateau few meters apart are likely to have similar pedons. Conventionally, *classification* is one of the methods in predicting values at un-sampled locations (from class average), but in this process a lot of information is likely to get lost; also the spatial association between points is not considered. Moreover, we are not usually able to measure the values of a particular phenomenon in all points of the study area, but only in sample points. Even if we are able to do so, it is not an easy task for a computer to analyze the seemingly endless data. Thus, we require a model that allows us to use the samples taken at some points only. The interpolation gives us values in such points where we have no measurements. At the same time, the method must also take full account of the way by which the attribute varies in space. Interpolation is just this technique.

The goodness of an interpolation technique can be evaluated by the discrepancy of the interpolated values from the true value. Because the true value is not known in general, we may select some measured points as 'true to the best' for testing the interpolation procedure. We may often have different attributive data in different (regular) spots. For example, we have two phenomena, temperature and precipitation, (averaged and summed up for a month, respectively), given in two grids, not coinciding. To perform the analysis we need to transform the data to a common grid (i.e., to a common coordinate system). Seemingly similar tasks of interpolation are used for the creation of area objects representing a particular interval of the data in question.

7.2 Spatial Interpolation

As discussed earlier, spatial interpolation refers to the procedure of estimating the values of attributes at un-sampled sites from measurements taken at point locations within the area covered by existing observations (the same process while performed for sites outside the specified area is termed as *extrapolation*). We can even say that the interpolation in a GIS has its role in filling the gaps between observed data points, which are otherwise stratified (regularly spaced observations but not at every point); or patchy (cluster of observations at specified locations); or random (randomly spaced observations). It may be used in GIS to provide contours for displaying data graphically (arithmetic map), to calculate some parameter of the soil surface at a given point, to change the unit of comparison when using different data structures in different layers (such as using satellite imagery and other map data together) and also as an aid in the spatial decision making process.

7.2.1 When is interpolation necessary

The followings are the examples when interpolation is necessary while doing the analysis in a GIS environment: (a) When an image viz. satellite image or any scanned documents of a specific grid-size is converted to an image of another size of *tessellation* (a mosaic typically consisting of small square tones) — the process is known as *convolution*; (b) when a continuous surface of one kind of tessellation needs to be transformed to another kind (TIN to raster or vector to raster); and finally (c) when a sample data sets are to be converted to continuous surface. The data structures for the continuous surface may be regular grids (raster), contour lines (isopleths) or triangulated irregular network (TIN) etc., but these data structure must provide value for each individual grid cell and hence space-saving data structures (like run-length codes and quadtrees) can not be used.

The final product can be displayed as maps or may be used for further analyses. In image-output, the values of the attributes are represented by zones of color or grey levels. Line-maps include isolines (lines of same value) or vertical slices (profiles) or critical lines like ridges, stream courses, shorelines etc. Here is a point to remember, the interpolated surfaces are not 3D representations. Though these surfaces might have the third axis as the interpolated attribute (distributed throughout an area defined by x, y coordinates), but these attributes do not vary continuously as seen in a 3D frame. Continuous volumes are usually represented by fence diagrams or by 3D display of a crisped image of a zone of concentration.

7.3 Methods of Interpolation

There are a number of methods of interpolation, which, depending on the objectives and the data set, should be chosen accordingly. Here we are making an attempt to classify them as follows.

7.3.1 Point/Areal interpolation

In point-based interpolation, we use some points whose locations and values are known to compute the values of points at other locations. This is used for data which are collected at point locations e.g., weather station data, spot heights, ground water samples etc. Interpolated grid points can be used to draw isolines (e.g., contours) using a linear interpolation between each pair of grid points through computer contouring algorithms. In case of area based interpolation, a set of data mapped on one set of source zones is used to determine the values of the data for a different set of target zones, e.g. populations for electoral districts may be estimated from the population counts for census tracts. Point to point interpolation is the most frequently performed type of spatial interpolation done in GIS.

7.3.2 Global/local methods

In global interpolation, a single mathematical function is used to map across the whole area of interest. This produces smooth statistical surface with minimum sudden changes. But a change in one input value affects the entire map. Local methods, in contrast, apply the mathematical function(s) to small subsets of the total data points to create local surfaces and link these surfaces to generate a composite surface for the whole area. In local interpolation method, a change only affects the result within the window without disturbing other sets. The distinction between a global and a local interpolator is very loosely defined as the local interpolator can be extended to include a large proportion of the data points making them (in a sense) global.

7.3.3 Exact/approximate

Exact methods of interpolation take account of all the observed data points. None of the data points are allowed to be missed or not taken care of. The surface thus produced, pass exactly through all points whose values are known. Proximal interpolators, B-splines and kriging methods take care of all the data points. On the contrary, approximate methods do not take all the available points but smoothen or alter some of the points to get a general trend. Sometimes in many data sets the global trends vary slowly; but the local fluctuations vary rapidly and may produce uncertainty (error) in the recorded values. In order to pay importance to the global trends, some modifications in some of the points are done to reduce the effects of error on the resulting surface.

7.3.4 Gradual/abrupt

Gradual methods produce a smooth continuous profile like a flat terrain, a typical example is the distance weighted moving average. This usually produces an interpolated surface with gradual changes. The abrupt methods are suitable where some discontinuities like ridges or valleys on the surface are expected or to include barriers in the interpolation process.

7.3.5 Deterministic/stochastic

When sufficient information is available about the variables and parameters for modelling geographical surface, one can go for deterministic models. But there is always uncertainty as found in the real world. Hence stochastic models are used, which assign the concept of probability distributions to some or all the variables and parameters. Stochastic interpolators include trend surface analysis, fourier analysis and kriging. This provides a powerful tool in solving the problem of uncertainties, where the final output is the most important.

However, irrespective of the kinds of methods, interpolation always provides an estimation of the values of attribute at un-sampled location based on the available observed points. Thus, there is always some degree of inexactness (whatever be the accuracy of the measurements and sampling) attached to the output. It is always advisable to keep in mind of this limitation of interpolation while doing further analysis in GIS.

As we have seen, there are different ways of classifying the interpolation methods. Keeping in view of the use by most of the people working with GIS, we will discuss the following groups of methods:

1. Method of geometrical nearness
2. Statistical methods based on weighted average
3. Methods using basis functions
4. Method of artificial neural networks

7.4 Methods Based on Geometrical Nearness

When spatial data are sparse, we assume that the observations are taken from a statistically stationary population, where mean and variance of the data are independent of both the location and the size of the support. *Support* is a technical term used in geostatistics for the area or volume of the physical sample on which the measurements are taken. If the weight of the soil sample collected from a soil profile kit were 1 kg, the support would be approximately 10×10 cm in area and about 5 cm thick. For a liter of ground water sample extracted from a sampling tube, the support would be the dimensions of the column. As the samples are homogenized in the laboratory all internal variation is lost. Hence we usually refer to the area of volume of the material sampled and not to a larger or smaller volume or area of sampling. For remote sensing data the pixel area is the support. Here, the spatial prediction is made through some classificatory approach. We can get out prediction by using well-known standard analysis of variance (Anova) methods. The simplest statistical model is the Anova model:

$$z(x_0) = \mu + \alpha_k + \varepsilon$$

where, $z(x_0)$ is the expected value of the attribute *t* at location $z(x_0)$, μ is the mean of attribute '*z*' over the area of interest, and a*k* is the deviation between μ and the mean of unit *k*; ε is residual (pooled within-unit) error (known as 'noise').

This model assumes that for each class *k* the attribute values are normally distributed; ideally, each contains a distinct mode. The mean attribute value per class *k* is $\mu + \alpha$, which is estimated from a set of independent samples that are assumed to be spatially independent. The mean within-class variance is given by ε, and is assumed to be the same for all classes.

The relative variance $\left(\sigma_w^2 / \sigma_t^2\right)$, where σ_w^2 is the pooled within-class variance and σ_t^2 the total sample variance, is a measure of goodness of fit for the classification. Both can be estimated when all map units contain more than one point sample. The lower the relative variance the better the classification.

The Method As a consequence of involvement in research on geometric aspects of crystallography, the Russian mathematician, Voronoy Georgy Fedoszeevich (1868-1908) discovered a special continuous subdivision of the space by convex polyeders. In two-dimension, the polyeders are turned into polygons called Voronoy cells. The geometric rule defining the cell is very simple. Each cell has only one sample point and all other points inside the cell are closer to this sample point than to any other sample outside the cell. These cells are also popularly known as Thiessen polygons (Fig. 7.1). To construct the Thiessen polygon (or Voronoy cell) for a sample point we have to connect all data points to the selected one and to drop the bisecting perpendiculars to each link (connect). The bisectors thus intersect with each other. Now we can select on each bisector the nearest to the sample point two intersections, determining the cell (Laurini and Thompson, 1992). Thiessen polygons can also be generated using distance operator as shown in Fig. 7.2. This creates the polygon boundaries as the intersections of radial expansions from the observation points. If the data points lie on a regular square grid, then the polygons are all equal, regular cells with sides equal to the grid spacing. If the data are spaced irregularly, the polygons also become irregular.

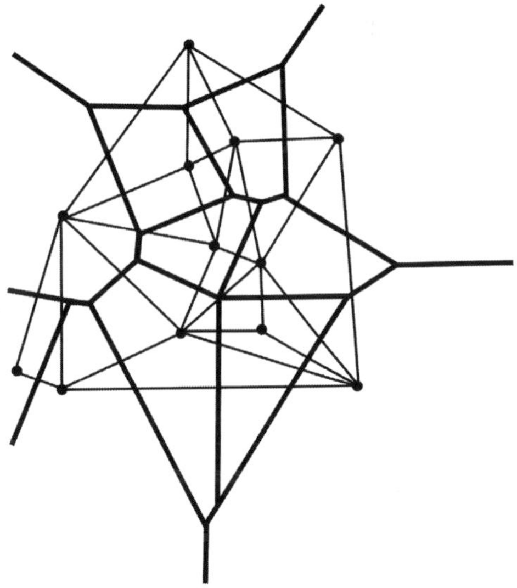

Figure 7.1: Thiessen polygon (thick lines) and Delauney triangulation (thin lines); • Data points.

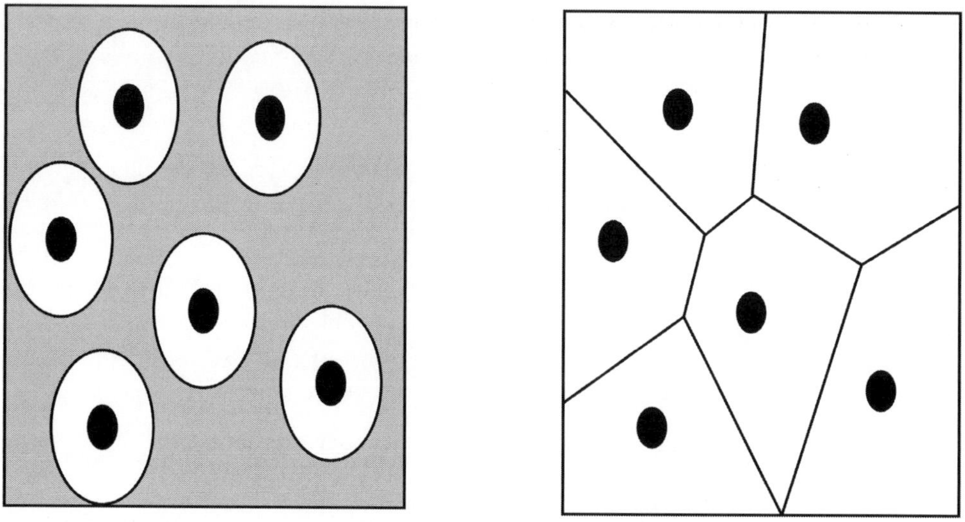

Figure 7.2: Generation of thiessen polygons (a) distance operator (b) thiessen polygons

Thiessen polygon method is a kind of point interpolation which is used when the sampled points are not densely located or have limited influence or continuity in surrounding observations, like weather station data. But every geographical phenomenon varies over the space and is not necessarily discrete. Its use is therefore limited. The *Voronoy* approach is suitable for dynamic spatial object condensation (Yang and Gold, 1995) i.e., for realization of nested GIS implementation models that make the storage and processing of the objects possible on different levels of generalization. Another application, the *Delaunay triangulation* (after Delaunay Boris Nikolaevich, Professor of Mathematics at the Moscow University, 1930), is related to constructing a continuous triangulation network of the measured values at sample points. If we use the links belonging to the borderlines of the cells (thin lines in Fig. 7.1), then this triangulation is established. On the basis of this the TIN modules of the different GIS software can be constructed. Delaunay triangulation is an 'exact' method as all the data points are taken into consideration. This is also a 'local' interpolator because the global characteristics of the data set have no influence over the interpolation process. And again, this method is abrupt as sharp boundaries are present between the interpolated polygons.

Triangulation method can easily be used with the quantitative data like vegetation or land use classes to generate the choropleth map (value-by-area-map). Here we assume that the variation within an entity (polygon) is nil and the change only occurs at the boundaries. Thus we get uniform areas of land use, which is definitely an approximation but very useful in practice. Tobler (1979) tried to improve this by introducing a continuous, interpolator that removes the abrupt changes due to inappropriate boundaries. His method was basically aimed to solve the problem of population distribution in administrative areas as people are usually not uniformly distributed, but tend to congregate in certain parts of the administrative unit. The method, known as *pycnophylactic interpolation* represents the distribution of people by a smooth continuous surface, which shows that the density of people in some parts of the area is higher, and in other parts lower, than the average. This method can be applied to any data by converting the data into a density function i.e., volume of attribute (here, number of people) per unit of spatial entity (here administrative unit). It is not an exact interpolator, as the values interpolated for individual grid cells differ from the original measurements.

7.5 Statistical Methods

The interpolation by statistical methods predict the function value in the unknown point using weighted average of the known values in the sample points:

$$F^*(x_0) = \sum_{i=1}^{n} \lambda_{i,o} F(x_i)$$

where $F^*(x_0)$ is the interpolated value in the location x_0, $F(x_i)(i = 1,2,....,n)$ is the value of the measurement in the sample point and the numbers $\lambda_{i'0}$ are the assigned weights.

The statistical methods must be local, because they involve only a number of neighboring samples (n in the above equation) located inside a distance limit (e.g., range in the kriging) from the point to be interpolated. All statistical methods smoothen out the data reducing the sudden juttings and dips but can interpolate only values that are inside the interval between the largest and the smallest sample values involved in the average building (Sárközy and Gáspár, 1992).

7.5.1 Interpolation by Kriging

Interpolation based on spatial dependence of attributes was first used by D.G. Krige (1956, 1960) in gold mining industry, which was generalized and extended to the "theory of regionalized variables" by Georges Matheron (1963, 1971). This theory led to the development of various procedures to estimate the spatially dependent variable, collectively known as *Geostatistics* — a very special branch of statistics.

Kriging refers to a method of interpolation based on the theory of regionalized variable, which makes optimal, unbiased estimates of the unknown values with minimum variances if the measured data fulfill conditions of stationarity (1st and 2nd order stationarity or intrinsic hypothesis). Kriging is based on the idea that the variable (attribute) change with some apparent continuity but can be modeled using a single mathematical function (DeMers, 2000). Kriging treats the statistical surface to be changing as a general trend (drift or structure of the surface) having some small variations from the general trend (spatial autocorrelation) and some *noise* (neither associated with general trend nor spatial autocorrelation). A beautiful explanation of these can be found in Clarke (1990).

In the case with no trend i.e. data stationary or data following intrinsic hypothesis (the expected difference of attribute values between two sample locations depends on the distance between them and not on their location), we use *ordinary* kriging. But when trend/drift is expected for the data (with normal distribution), *universal* kriging is to be used. The basis of this technique is the rate at which the variance between points changes over space. This change or drift is estimated using a mathematical equation fitting to a trend surface. This is expressed in a statistical graphing technique called semivariogram (Fig. 7.3), which is displayed as a plot of semivariance $\gamma(h)$ versus distance between points $\gamma(h)$, also called lag.

The semivariance statistic (*h*) can be defined by:

$$\gamma(h) = E[z(x)-z(x+h)]$$

When there are *n*(h) pair of observations, the semivariance at a given lag is estimated as the average of the squared differences between all observations as:

$$\gamma(h) = 1/2N(h)\sum_{i=1}^{n}[z(x_i + h)]^2$$

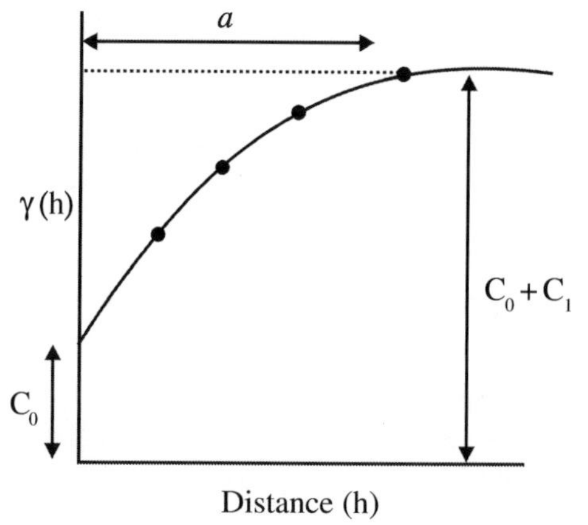

Figure 7.3: An ideal semivariogram model

Most semi-variograms show behavior as shown in Figure 7.3. Semivariance increases with distance between sample locations. The upper limit (asymptote) of $\gamma(h)$ is called the *sill* and the distance at which this limit is reached is called the *range of spatial dependence or simply range* (*a*). The sill approximates the sample variance s^2 for stationary data. When the distance between two observations is less than *a*, the observations are specially related. Range also defines the maximum radius from which the neighbouring samples are drawn through interpolation by kriging. It is to be noted that the fitted curve does not run directly through the origin, rather it intersects with the y-axis. This intersection is an estimate of the residual, spatially uncorrelated variance called *noise* or the nugget variance (C_0). This nugget variance represents unexplained or random variance often caused by the measurement error plus the microvariability of the attribute that can be detected at the scale of sampling (Burrough, 1986). Increase in nugget variance implies increase in measurement error or short-range random variation. In case of pure nugget effect (~ 100% sill), the variation is short-range and there is no special dependence. C_1 represents the structural variance and the some of C_0 and C_1 almost equals the sill.

There are various kriging procedures and a detailed account of the theory and applications are beyond the scope of this book. Here we are listing some of the common kriging procedures:

1. Punctual kriging: This is based on simple point estimation assuming data are collected at equally spaced locations and are isotropic in nature.
2. Block kriging: When the unit or information is management unit i.e., discrete areas or blocks.
3. Co-kriging: This is used when a variable (primary) cannot be sampled because of some difficulty and other close related feature (secondary) are sampled e.g. heights – secondary, rain – primary. Here, cross semivariogram is used to estimate the co-regionalization of these two variables.
4. Universal kriging: As already discussed, this type of kriging estimates surface from irregularly spaced samples when trend exits (condition of non-stationarity). This method is developed to permit kriging in presence of strong trends.
5. Kriging of log-normally distributed data. This includes log normal kriging (for data distributed log-normally) and disjunctive kriging (for variable with complex probability distribution).

Once the semivariogram is developed, it is used to estimate distance weights for interpolation. Interpolated values are the sum of the weighted values of some number of known points where weights depend on the distance between the interpolated and known points. The problems associated with this method are that when number of data points is large, kriging becomes computationally very intensive and the estimation of the variogram is not simple. Again there are assumptions that must be made about the statistical nature of the variation and the results from this technique may not be absolute.

7.5.2 INVERSE DISTANCE INTERPOLATION

It is the most common method of interpolation in GIS. Inverse distance interpolation integrates the concept of Thiessen polygons and the gradual change of trend surface. This method is based on the idea of positive spatial autocorrelation and it assumes that the value of attribute z at some unsampled point is a distance-weighted average of the neighboring data points within a user-defined specified area (window). Thus the calculated value for a location is based on the range of values of neighborhood points that fall within the window, and a function of the distance and the magnitude of the surrounding points. Points closer to the unknown point are given more weightage than those far away. The *inverse* distance weights are used to attenuate the influence of the distant points. To select the neighborhood (window) one can use a number of search techniques i.e., defining neighborhood as a circle of some predefined radius, selecting number of data points or dividing the number of points into quadrants or octants and selecting accordingly. The weighing factor may be square or cube or even higher power of the distance or some factors to modify the distance. And in case of developing surface with barriers, one should be much careful as these objects do affect the output of interpolation to a great extent.

Moving Average/Distance Weighted Average This method estimates the average of values at selected number of known points (Fig. 7.4):

In the figure, for $i = 1$ to n, the average value

$$Z = \Sigma w_i z_i / \Sigma w_i$$

where w is some function of distance.

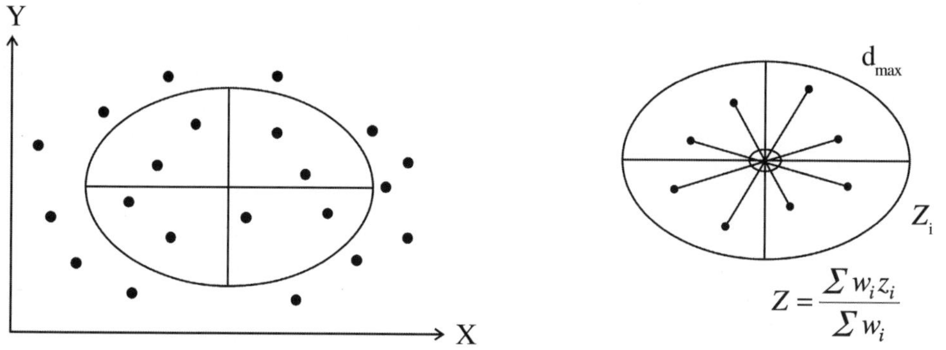

(a) A window for point selection **(b) Distance to observation points**

Figure 7.4: Principles of weighted averages

Here, a number of algorithms may be used, depending on the nature of the distance function, number of points used, and the direction from which they are selected. It has limitation in the range of interpolated values, as no interpolated value will be outside the observed range of known values. These are confusions in selecting number of points to calculate the average, as well as to deal with the irregularly spaced points and edge effects. Poor choice of window can affect the results badly, specially with the highly dense data.

7.6 Methods Using Basis Functions

There are methods like polynomial interpolation, splines with tension using radial basis functions, Fourier and wavelet transformations etc. that use a set of *basis functions* with the closest fit to the samples. Fourier analysis dissociates time series of spatial data into sets of sine and cosine waves, which are then overlaid to obtain the surface. Though Fourier analysis is the best for data sets that exhibit marked periodicity and the computation is relatively simple, it is rarely incorporated in GIS packages. The numerical methods of transformation and inverse transformation however work for regularly located samples but this form is unusable for interpolation of scattered points. Thus wavelet interpolation is still at its infancy in the solution of spatial interpolation problems.

7.6.1 Trend Surface Analysis

If an attribute varies continuously in space (as is generally observed in nature), the variation can be effectively modeled by a smooth mathematical surface. The surface is approximated by a polynomial equation involving the observed data and the values of attributes at the unsampled location can be predicted from their coordinates. This means that the output data structure is a polynomial function, which can be used to estimate values of grid points on a raster or the value at any location. A trend surface is a 'global' interpolator that assumes the general trend of the surface is more important than random minor errors that may be found at each sampled point. The global interpolation can also be made on different resolution levels. The choice of these depends on the task to be performed and the behavior of the sample itself.

The trend surface can be smooth or flat or undulating depending on the degree of complexity involved. Accordingly they are approximated by first order, second, third, ….., ……, n-th order polynomials. The simpler the trend surface, the lower will be the degree or order. By choosing the order of the surface one should take into the consideration the rule of thumb any cross-section of a surface of order n can have at most $(n\text{-}1)$ alternating maxima and minima. If applied to the terrain one can see the main characteristics of the surface in advance and accordingly choose the proper order. However the surface of other phenomena (e.g. temperature) is not visible and one should make several attempts to find the appropriate order.

The long-range spatial variation can simply be modeled by constructing a multiple regression equation of values of the attributes vs. their geographical location and to fit the polynomial line or surface (depending on whether our data is one-dimensional or two-dimensional) to the sample data using the least square adjustment. Thus the square of difference of the given and interpolated values is minimized (i.e., $Z(x_i) - z(x_i)$ is minimum, where $Z(x)$ and $z(x)$ are the interpolated and the measured values of attribute z at location x_i).

Considering z having measured at locations $x_1, x_2, x_3, \ldots, \ldots, x_n$ increase linearly with location, x, long-range variation of z can be approximated by the regression model

$$z(x) = a_0 + a_1 x + \varepsilon$$

where, a_0 and a_1 are the polynomial coefficients known as the intercept and the slope respectively in regression. The residual ε (the noise) is assumed to be normally distributed and independent of x.

The quadratic or higher order of polynomial models may be used if z varies non-linearly with x.

$$z(x) = a_0 + a_1 x_2 + \varepsilon$$

ε reduces to almost zero for an higher order equation. For surface, the polynomials take the form

$$f\{(x, y)\} = \sum_{r+s \leq p} (b_{rs} \cdot x^r \cdot y^s)$$

where x and y denotes the coordinates, r and s are any number and p is the order of the trend surface. There are $P = (p + 1)(p + 2)/2$ coefficients that are normally chosen to minimize

where, the *i*-th sample value is denoted by $f(x_i)$, the interpolated value in the same point by $z(x_i)$, and the number of samples is *n*.

The value of the minimum depends on the suitability of the chosen model for the particular data set. As a measure of the goodness of fit we can use the parameter called 'mean square error' (MSE).

As the above, the flexibility of the polynomial depends on its order. A linear equation (first order) describes a tilted plane surface: $z = a + bx + cy$; while a quadratic equation (second order) describes a simple hill or valley: $z = a + bx + cy + dx^2 + exy + fy^2$; and the equation for the cubic surface (third-order) will be: $z = a + bx + cy + dx^2 + exy + fy^2 + gx^3 + hx^2y + ixy^2 + jy^3$ and so on.

The trend surface interpolator can also be used and so on for three-dimensional modeling. In this case it is called as *polynomial regression* (Collins and Bolstad 1996). The authors used for the estimation of the temperature *t* by following the expression, $t = b + b_1 x + b_2 y + b_3 x^2 + b_4 y^2 + b_5 xy + b_6 x^3 + b_7 y^3 + b_8 x^2y + b_9 xy^2 + b_{10} z$.

In their opinion the difference between the trend surface analysis and the polynomial regression are the number of independent variables (the trend surface as a surface can have only two variables) and the completeness of polynomial expansion (the trend surface analysis has complete expansion, the polynomial regression can be set up using the linear combination of independent functions). We can see, that in the formula of the temperature the variable *z* is applied only in linear form.

Trend surface analysis is easy to understand the way by which the surface is calculated. Broad features of the data can be easily modeled by low order trend surfaces. However, for higher polynomial order more numerical problems may occur by risk of emerging huge, ill-conditioned matrices, which should be inverted. As a conclusion, for most of the practical tasks the order should not exceed five. One problem with this method is that the statistical assumption of the model is rarely met in practice. Another problem might be that the edge effects may be severe. The polynomial trend surface has the unpleasant behavior to increase or decrease very rapidly in the places where the density of samples is rare especially on the boarders of the region. Second order or higher surfaces may yield ridiculously large or small values just outside the area covered by the datasets. Again, one of the very practical uses of trend surface analysis, as pointed out by Burrough et. al (1977) and Davis (1986) is to reveal parts of a study area that show the greatest deviation from a general trend. This then could be used to remove the broad features of the data prior to using some other local interpolator.

7.6.2 Splines

Before the arrival of computer the smooth curve fitting for the data points were performed using flexible rulers. These rulers, called 'splines' were held in place by weights on pegs at the data points while the line was drawn (Pavlidis, 1982). Today this ruler is placed by a third order polynomial, which is applied between two adjacent points under the condition that the first and second order differentials should be continuous. Such a curve is called *spline* (Fig. 7.5). The polynomial function is given by $y = ax_3 + bx_2 + cx + d$

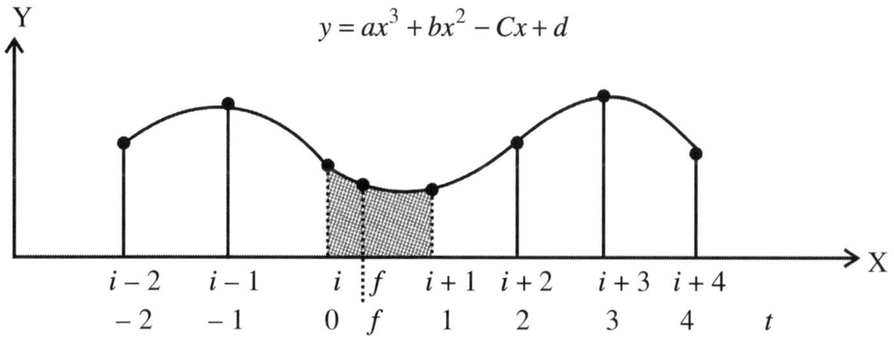

$$y = y_{i-1}(-t^3 + 2t^2 - t)/2 + y_i(3t^3 - 5t^2 + 2)/2$$
$$+ y_{i+1}(-3t^3 + 4t^2 - t)/2 + y_{i+2}(t^3 - t^2)/2$$
$$0 \le t \le 1$$

Figure 7.5: Spline

The points that divide an interval *i*-2, *i*-1, *i*, *i*+1, *i*+2 etc., are called *break-points* and the points of the curve at these values are commonly called *knots*. This is also called cubic interpolation. The term *spline*, was first used for cubic piece-wise polynomial functions, hence it is also known as the *cubic interpolation*. The term *bicubic spline* is used for the three-dimensional situation where surfaces instead of lines need to be interpolated. In case, when the curve is not a single function of *x* as shown in Figure 7.6, an auxiliary variable *u* should be introduced as follows:

$$x = f(u) = a_2u^3 + b_2u^2 + c_2u + d^2$$
$$y = g(u) = a_2u^3 + b_2u^2 + c_2u + d_2$$

Splines are fitted to a small number of data points exactly and also ensure that the joining between two parts of the curve is continuous. This ensures continuity in (i) elevation (zero-order continuity) — surface has no cliffs; (ii) slope (first-order continuity) — slopes do not change abruptly, and (iii) curvature (second order continuity) — minimum curvature is expected. Spline produces a continuous surface with minimum curvature and the output data structure is points on a raster. Some of its advantages are:

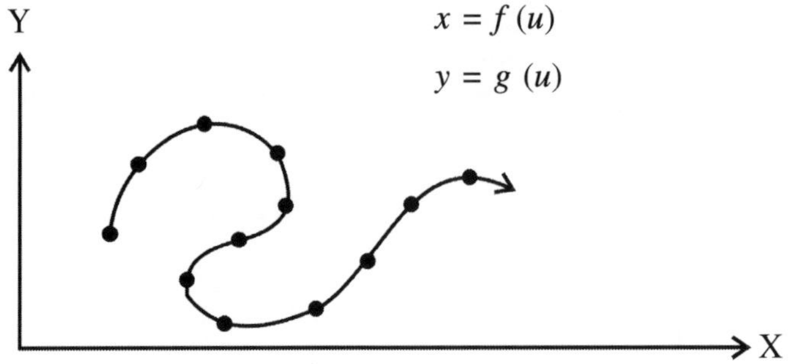

Figure 7.6: Spline with auxiliary variable (u)

1. It is possible to modify one part of the curve without having to recompute the whole, which is not possible with inverse distance interpolation. Hence adjustments are made locally;
2. Can be exact or used to smooth surfaces. Exact splines round off the sharp corners;
3. Computing load is moderate.

Spline is best for very smooth surfaces, but poor for surfaces, which show marked fluctuations. The later can cause wild oscillations in the spline. Thus splines are popular in general surface interpolation packages but are not very common in GIS packages.

In practice, it is very difficult to calculate simple splines over a wide range of separate sub-intervals (say for digitized lines). A special kind of spline called B-spline is used here. A cubic curve is determined by using four adjacent observations as shown. B-splines are used for smoothening digitized lines in most GIS at the boundaries. But this process of using B-splines in smoothing-off the boundaries may create error in computing the perimeter or the area. The area as calculated using the digitized data points (accurately done) will be different from that using smooth boundary. The sinuous boundaries that also include sharp, rectangular corners should not be smoothened by using B-splines. Exact splines are used to smooth the curves and contours. But for two-dimensional or three-dimensional data, exact splines have limitations. In this case, thin plates splines are used which employ a locally smoothed average surface in place of the exact spline.

7.7 Artificial Neural Networks

And now, a few comments on this very new paradigm of science which, though well established in many areas, is new in the domain of spatial interpolation and GIS. The Artificial Neural Network (ANN), an information processing paradigm is inspired by the way the biological neural (nervous) systems (such as brain) process information. Though human of biological neural networks are much complicated and highly interconnected, the mathematical model so used in ANN are relatively simple and are configured for a specific application, be it pattern recognition or data classification or trend analysis. An ANN resembles the human brain in two aspects (Haykin, 1994) viz. knowledge is acquired by the network through a learning process and synaptic weights (i.e., the strengths of the interneuron connection) are used to store the knowledge. The theory and basic algorithms of ANN were developed in the last decade but the software products actually arrived only during '90s' and the applications for spatial interpolation started recently.

A simple MLP (Multi-Layer Perception) network has three principally different types of layers: an *input layer*, *hidden layers* (any number), and an *output* layer. The number of nodes in the input layer corresponds to the number of independent variable of the particular task, while the number of nodes in the output layer depends on the number of scalar functions involved in the common evaluation, the components of a vector. The network topology is *feed-forward* and each node must be connected to another node of the next forward layers. The nodes of hidden layers and output layers contain activation functions, which are mostly nonlinear (sigmoid or hyperbolic tangent or the Gaussian bill-curve). The procedure to adjust the weights to fit the given data is called the *training* or *learning*. The network is *trained* (say, for interpolation) by learning from known input-outputs couples of data. Once the training is performed properly, and then only variables at unknown points can be applied as input to generate the output. In order to obtain the output close to desired output, some backward pass are performed to adjust the network's (neurons') weights. This process is called *back propagation*. Through back propagation, the error on the output (called error function) is characterized to a minimum level.

ANNs may be applied to problems which are much complex to deal using conventional techniques. These include problems that either have no algorithmic solution or the developed algorithm is too complex. Neural networks help where we are able to perceive a lot of examples of the behavior we require or where we need to pick out the structure from existing data. A great deal of research is going in neural network worldwide. There are already some successful stories of applications of ANN in business world. In GIS, and more specifically in interpolation, applications are coming up. ANN was able to perform well in interpolation of precipitation values in contrast to other methods like Voronoy cells. Inverse distance weighted and ordinary kriging (Xingong, 1997). A non-linear ANN model was found to provide better representation of rainfall-runoff relationship of a river basin than using time series approach (Hsu et al., 1995). Artificial neural network also has been widely used as models for a variety of other hydrology processes. There are studies that deal with the use of neural networks connecting the object oriented spatial data model with the function field data model (Pariente 1994; Sarkozy 1994).

The neural networks have also been use extensively in classification issues like processing of multispectral remote sensing scenes, pattern extraction from images and also for classification of land quality in GIS (Hepner et al., 1990; Yoshida and Omatu, 1994; Muttiah et al., 1996). There are other specific examples and to deal with all of them is beyond the scope of this book. The last point to say that there has been a tremendous attention and funding in research in the field of ANN. Future research in many fields are likely to come and certainly, GIS will play a key role there.

CHAPTER 8

OUTPUT

A map is the greatest of all epic poems. Its lines and colors show the realization of great dreams.”

Gilbert H. Grosvenor, Editor of National Geographic, 1903-1954

Technology is meant for improving our life, and GIS is no exception. We have already seen that the decision-making processes are driven by the results of analyses through GIS. An ‘output’ from a GIS may lead to wrong decisions, if not presented in a proper way (leaving aside issues of data quality, potential errors in analysis etc.). Output must also be as simple as possible and should not leave any ambiguity. Though a GIS output is referred to describe the ways in which information from a GIS can be presented through a GIS (Heywood et al., 2000), it is not an isolated function of GIS and more so, because an output from GIS does often play as an input to the decision making or problem solving process. A reliable output is thus a product of chains of activities starting from data through analysis and finally the production of the analyzed data; each step influencing the decision making through GIS.

There are several forms of output like text (tables, lists, numbers or text in response to query), graphics (maps, screen displays, graphs, perspective plots), digital data (on disk, tape or transmitted across a network), or computer-generated sound and images. But since maps (graphic output) are the most common visualization of GIS output, let us discuss it first.

8.1 GRAPHIC OUTPUT

Among the graphic outputs, the most common from is the map. Map provides spatial information including size, shape, location, pattern, trends etc. According to definition given by International Cartographic Association, the map is ‘a representation, normally to scale and on a flat medium, of a selection of material or abstract features on or in relation to the surface of the Earth’. A map is a spatial model of the real world, but differentiated by certain key elements important to achieve the best of the objectives of mapping, as listed by Robinson et al. (1995) as the frame of references, the projection used, the features to be mapped, the level of generalization, annotation used, and the symbolism employed. We can have here to more important additions here—the purpose and the simplification. Maps may be found of many types, viz. planimetric, topographic, cadastral, image or thematic, each made with a specific purpose. The planimetric maps are designed to portray the horizontal positions of features while the vertical information is ignored, whereas a topographic map portrays features on the surface of the Earth, including relief (elevation), hydrography, and even, cultural features. A cadastral map represents boundaries of land parcels, ownership, land use etc. An image map is usually a photograph taken from an aircraft or a satellite that is a reflection of all or part of the Earth's surface. But the most common form of map is thematic which is used to visualize spatial relationships and patterns among information pertaining to some theme or concept (e.g., soil, land use, geology, population density maps etc.). Thematic maps can again be classified as choropleth map (uses zones or polygons to display information using shading, dot, density, or other techniques, e.g. population change, ethnicity distribution), *proximal* (*dissymmetric*) map (shows zones of constant attributes, such as soil type of vegetation e.g., zoning, soil map, *isopleth map contour or isarithmic* (shows a continuous three-dimensional surface such as elevation using lines connecting points of equal value e.g., elevation, land values, income) or *point* (*dot*) *or symbol map* (shows information relating to specific points using marker symbols whose size and/or frequency relates to magnitude of phenomena e.g. housing sales, code violations, crimes).

Expressing the spatial information in map form is probably the most elegant, compact and powerful method of visualizing and displaying data, even today (Heywood et al, 2000). The map design elements are subjected to human perception and thus vary widely. Thus one has to look into the science of Cartograph to find out the principles of map design, which itself is too vast to come within the scope of the book. We will, therefore, discuss some of these principles in short leaving mostly to the choices of the user (see Box. 8.1).

Map design is as much an art as it is a science (Heywood et al. 2000). One must have a look at different authentic published maps to have an idea of the use of various graphic and non-graphic modes that are generally used to create a map. Nevertheless here are few points to note. A grid may be used to show the reference i-th latitudes and longitudes or planar (x, y) coordinates. This is useful to show clearly the scale and orientation of the map. One can also use an arrow showing the north to define the orientation and a graphical or numerical scale bar must be used to show the size of the map (reduction to the original size of real world features). An indicator map (small) to show the exact location of the main map area is also desirable. The prime objective of a map must be to communicate effectively and the scientific principles vis-à-vis artistic dignities play the crucial role.

To produce a map depicting accurately the real world features, one should use appropriate projection system also. For large scale maps of small areas, planar coordinates derived from a local projection system should be the choice while for maps of larger areas, decision have to be made about which is most important—area, distances or lines of true scale (Heywood et al. 2000).

Charts are mostly used to display tabular data. The types of charts available in most of the GIS are typical and very familiar to common use (e.g. area, bar, column, line, pie, and scatter with many forms). Charts should also be dynamic i.e., charts must automatically be updated as data in underlying table changes).

Box 8.1 Making Better Maps: Some General Principles

1. Prepare the map as simple as possible and never get bogged down into the software details.
2. It is wise to show only the area under study, without giving details of other areas.
3. One can insert small map to show relative location of the area.
4. Decide the scale or final printing. For report of general publication, reduction of desired size is essential.
5. Title must be short and to be put at the top using 'thick line' font.
6. Map labels are best when given in serif font (letters have 'tails'), with first letter upper and the rest lower case.
7. Map labels should be placed above and to the right of the feature. Alternately, it can be given above to left of the feature.
8. Font size can be used intelligently to indicate relative importance of labels.
9. Maps may be in black and white, colors should be added depending on the purpose and requirement.
10. While using colors use low saturation color in background, high saturation in foreground.
11. One can have choice of shading depending on relative quantity of data. Symbols should have visual progression corresponding to data values.
12. For quantitative data, polygons should have monochromatic color with different saturation or value of color, while for point symbols, different sizes of the same symbol should be used.
13. For qualitative data, use different colors (polychromatic) or different fill pattern for polygons; for point symbols, use same size of different symbols.
14. Legend should be big enough to show clearly different fill patterns.
15. Try to use the same theme twice in a map (legend should not be duplicate).
16. Use buffering to create special effects e.g. along a shoreline, create multiple buffers then color and ramp them.
17. For dot patterns density of dots should be kept constant and only the size of dots should be changed.
18. Use color or fill pattern for oceans and large water bodies. Other than maps and charts a very useful nor-cartographic output is linked display, which dynamically links various data display windows through a graphic user interface (GUI). Thus any action taken place in one data display

is automatically executed and visualized in another data window. Thus maps, charts or tables are linked under one system, thus providing the user a very useful and interactive mode of exploring spatial and non-spatial or attribute data. The linked display facility of a GIS provides accessibility to any data or application directly from a view using hot links. For example, one can clips (if already available) about that state. Thus photographs, images, texts, graphics and even audio and video all can be linked to map. This adds to the value of a map display and gives the user, ample scope to interact virtually at every possible step to make its decision perfect.

8.2 Text Output

Text forms are usually referred to as non-cartographic output, which is sometimes very important as in case of reporting results of analysis. Text includes list or table or charts of selected objects with attributes. Queries might result in numerical or character information e.g. totals, distances, areas or counts. Text output can also be delivered by a voice generator. Similar form is a chart, which displays attribute information linked to spatial entities (point, line or polygon) as numbers or character strings. Lists are of little value without an accompanying map to identify objects in the list.

8.3 Device Output

The device or mechanism to produce GIS output varies in a similar very as the forms of output vary, ranging from permanent to ephemeral or virtual types. The permanent devices make out permanent copies of maps, charts or diagrams (hard copy) that are either printed or plotted on paper. The ephemeral output on the other hand is mainly provided by the computer screen. The latter type is interactive in nature which allows users to be viewed for a time and then gets cleared from the screen. Television like terminals which came only in early 1960's, are now the most common way through which users interact with computer systems. Let us now have a brief look at devices for generating output and the ways their development influenced our life.

8.4 Hardcopy Output Mechanism

The hardcopy outputs may be maps or charts or other outputs that are printed or plotted on paper or film. There are a wide variety of media to deliver these types of outputs; some of the most common are discussed below.

8.4.1 Line printers

This was probably one of the first common devices for hard copy output, capable of printing 1,000-200 lines of text per minute, where each line is with 132 characters. The problem with the type of printers is the limited set of characters that can be printed. The image has to be created row by row from the top with fixed resolution. It is also difficult to draw continuous lines, and the outputs are best when viewed from a distance. The resolution of line printer is expressed in pixels or megapixels (~106 pixels) and is the most useful for repeated mapping of statistics on a constant base like census atlas, weekly reports of different statistical data.

8.4.2 Dot matrix printers/plotters

A very low cost device, dot matrix printers build image/picture with uniform pixels, usually in rows, 9 or 25 at a time. The dots are printed by impact of hammers on ribbons, but more recent versions use lasers and xerography technology, which can give high resolution (up to ~ 300 dpi). The color inkjet or bubble jet printers are also available that use an intelligent mixing of three or four primary colors. Though dot matrix is cheap, the quality of output is very low compared to line printers or inkjets. Shades of grey are generated from random dots that are printed in a small area.

8.4.3 Pen plotters

A pen moves under computer control and an image is created incrementally i.e. drawing a line using large numbers of movements of fixed size in fixed directions. These plotters usually have two motors; each for vertical and horizontal direction and the diagonal movement is in eight directions using combinations of one or both of the motors. From GIS perspective, an important advantage of pen plotter is the ability to plot on top of pre-printed base maps, thus avoiding all of the map information to be in digital form. There are many types of plotters depending on whether the paper is flat, or rolls over drum or pens, and can be changed automatically to generate different colors, or replacement of pens by tools or time taken for plotting jobs etc.

Similarly, there are electrostatic color plotters, thermal printers, laser printers and photographic image setters available in the market at varying costs, each producing different quality of outputs.

8.4.4 Optical scanners

Mostly used for remote sensing and other image processing, optical scanners generate output on photographic paper, which is mounted on inside of a rotating drum. The image is created by rotating drum and moving light source along the axis of drum. These are very costly device, but the quality and resolution of output are very high as compared to jet-printers or potters.

8.5 CRT (Cathode Ray Tubes)

Cathode ray tubes are the most common among screen types used for display of output, the other being the liquid crystal displays (LCD). The basic element of the image is pixel and so CRT is classified as raster device (contrary to this, plotters are vector devices).

When Tektronix introduced screens based on storage tube technology in 1970, it was a major breakthrough in lowcost graphic display images. Here images are generated by electron beam that moves over screen under control of computer. But the image is permanent and must be erased completely to produce another fresh one.

Storage tube technology soon got replaced by refresh image technology. Images can be refreshed by lighting dots in fixed positions. The resolution is determined by the number of rows and columns of dots, which vary from one graphic adaptor (GA) to another. Colors are created by illuminating three dots of primary colors— red, green and blue in different degrees.

An improvement in the ephemeral output began with the use of bit planes and palettes. These terminals (screens) use one bit of storage per pixels for black and white image and several bits for color display. The number of bits used per pixel gives the number of planes e.g. 4-bit planes, 20-bit planes etc., and the number of bit planes denote the number of colors used for display, which is different from the number of possible colors. Simultaneously the set of colors chosen for one time display is called the *palette.*

Another improvement in this aspect has been the three-dimensional stereo display, which is a very popular method for visualization of spatial information, specially the topographic data. These devices use orthographic projection based on angle of view (0 to 360 degrees), view azimuth (0 to 90 degrees from horizontal to vertical) and view distance (distance between viewer and image) (Heywood et al., 2000).

CHAPTER 9

REMOTE SENSING AND GIS

"Man must rise above the Earth to the top of the Universe and beyond, for only then will fully understand the World in which he lives."

Socrates, 500 BC

9.1 INTRODUCTION

Spatial information derived from remotely sensed imagery is often further analyzed within a GIS. Imaging remote sensing provides spatial data on electromagnetic radiation emitted or reflected by the Earth's surface. Through analysis, including digital image processing and techniques for information extraction, these data may be converted into information relevant to a GIS. There are reasons why GIS and remote sensing can benefit from mutual integration and these are (i) analysis of remote sensed imagery can benefit from GIS based data, (ii) remote sensing data can be the basis of up-to-date geo-information and (iii) integration of the information extraction process in GIS may help to keep track of errors and uncertainties inherent in the capture and manipulation of data. This chapter includes definition and principles of remote sensing, remote sensing as data source and several aspects of integration of remote sensing and GIS.

9.2 REMOTE SENSING

Remote sensing is the science of obtaining information about an object or area through the analysis of measurements made at a distance from the object (i.e., not coming in contact with it). In remote sensing, the sensors are not in direct contact with the objects or events being observed. The information needs a physical carrier to travel from the objects/events to the sensors through an intervening medium. The electromagnetic radiation is normally used as an information carrier. The quantity most frequently measured and recorded in images is the electromagnetic energy reflected by the object. The restriction to electromagnetic waves is due to the fact that the observation from a spacecraft excludes other possibilities such as sonic waves, which require a medium like air, water or solid earth for propagation. Other means of indirect observations for example, by stationary magnetic or electric fields are not sensitive enough for high geometric resolution. The output of a remote sensing system is usually an image representing the scene being observed. A further step of image analysis and interpretation is required in order to extract useful information from the image. The human visual system is an example of a remote sensing system in a general sense.

A remote sensing system that possesses only a sensor and depends on the external source to irradiate the target to be sensed is called a passive system. In general, Sun is the source of the electromagnetic energy in passive remote sensing system. But the active remote sensing system possesses both the sensor and the source of the signal. For example, LIDAR is an active system, which transmits signal from a LASER source to irradiate a target, and receives the LIDAR returns by a suitable light detector.

9.3 Principles

Remote sensing encompasses the set of sensors, platforms, and the data processing techniques that are used to derive information about the physical, chemical and biological properties of the Earth's surface features (land and ocean) and atmosphere using sensor onboard (air craft or balloons) or space borne (satellite, space shuttles) platforms (Fig. 9.1). The oldest form of remote sensing is aerial photography where the sensor system is the camera and film, but it has grown to include electro-optical sensors in optical or microwave regions of electromagnetic spectrum (Fig. 9.2). The acquired multi-spectral digital images are processed and analyzed by computers.

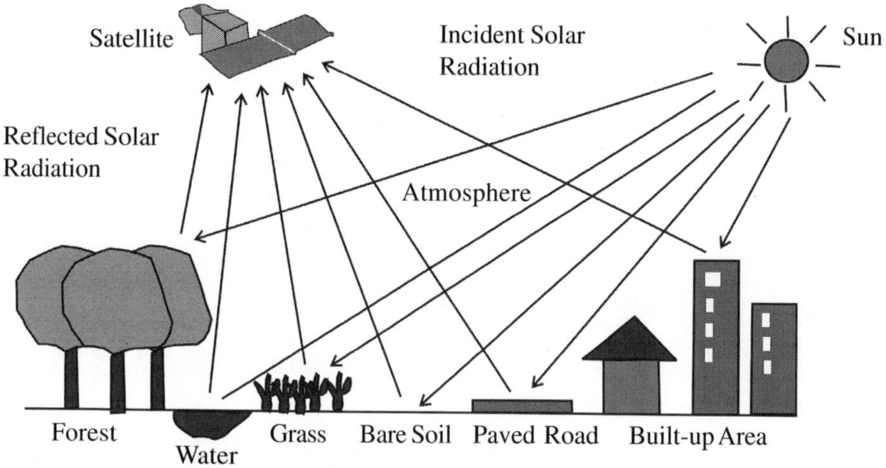

Figure 9.1: Principle of remote sensing

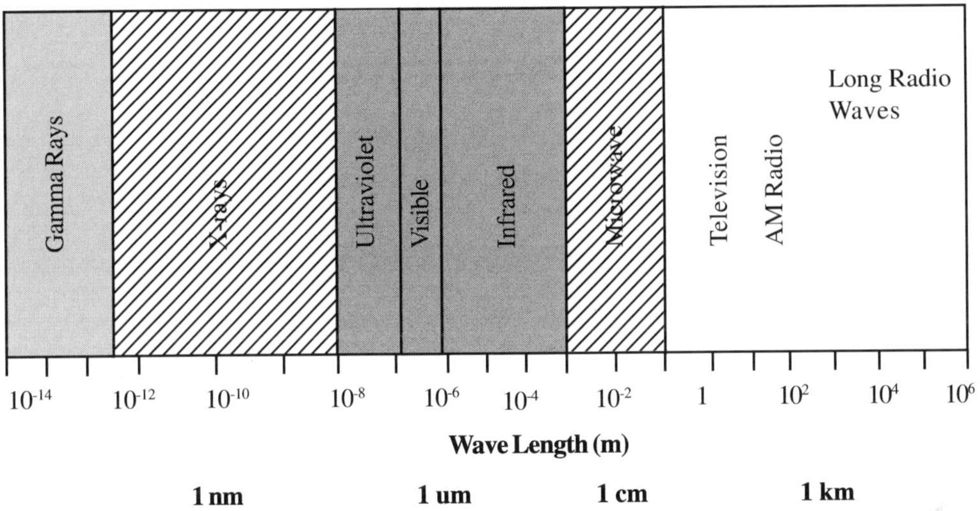

Figure 9.2: Electromagnetic spectrum showing regions used for remote sensing

In optical remote sensing, optical sensors detect solar radiation in the visible and near infrared wavelength regions reflected or scattered from the Earth, forming images resembling photographs taken by a camera high up in the space. Some remote sensing satellites carry passive or active microwave sensors. The active sensors emit pulses of microwave radiation to illuminate the areas to be imaged. Images of the Earth's surface are formed by measuring the microwave energy scattered by the ground or sea, back to the sensors. These satellites carry their own *flashlight* emitting microwaves to illuminate their targets. The images can thus be acquired day and night. Microwaves have an additional advantage as they can penetrate clouds. Images can be acquired even when there are clouds covering the Earth's surface.

A microwave imaging system, which can produce high resolution image of the Earth, is the synthetic aperture radar (SAR). The intensity in a SAR image depends on the amount of microwave backscattered by the target and received by the SAR antenna. Since the physical mechanisms responsible for this backscatter is different for microwave, compared to visible/ infrared radiation, the interpretation of SAR images requires the knowledge of how microwaves interact with the targets.

In satellite remote sensing of the Earth, the sensors are looking through a layer of atmosphere separating the sensors from the Earth's surface being observed. Hence, it is essential to understand the effects of atmosphere on the electromagnetic radiation traveling from the Earth to the sensor through the atmosphere. The atmospheric constituents cause wavelength dependent absorption and scattering of radiation. These effects degrade the quality of images. Some of the atmospheric effects can be corrected before the images are subjected to further analysis and interpretation.

A consequence of atmospheric absorption is that certain wavelength bands in the electromagnetic spectrum are strongly absorbed and effectively blocked by the atmosphere. The wavelength regions in the electromagnetic spectrum usable for remote sensing are determined by their ability to penetrate the atmosphere, which depends on its transmittance properties. These regions are known as the atmospheric windows. Remote sensing systems are often designed to operate within one or more of the atmospheric windows in optical and microwave regions of electromagnetic spectrum. These windows exist in the microwave region, some wavelength bands in the infrared, the entire visible region and part of the near ultraviolet regions (Fig. 9.3). Although the atmosphere is practically transparent to X-rays and γ rays, these radiations are not normally used in remote sensing of the Earth.

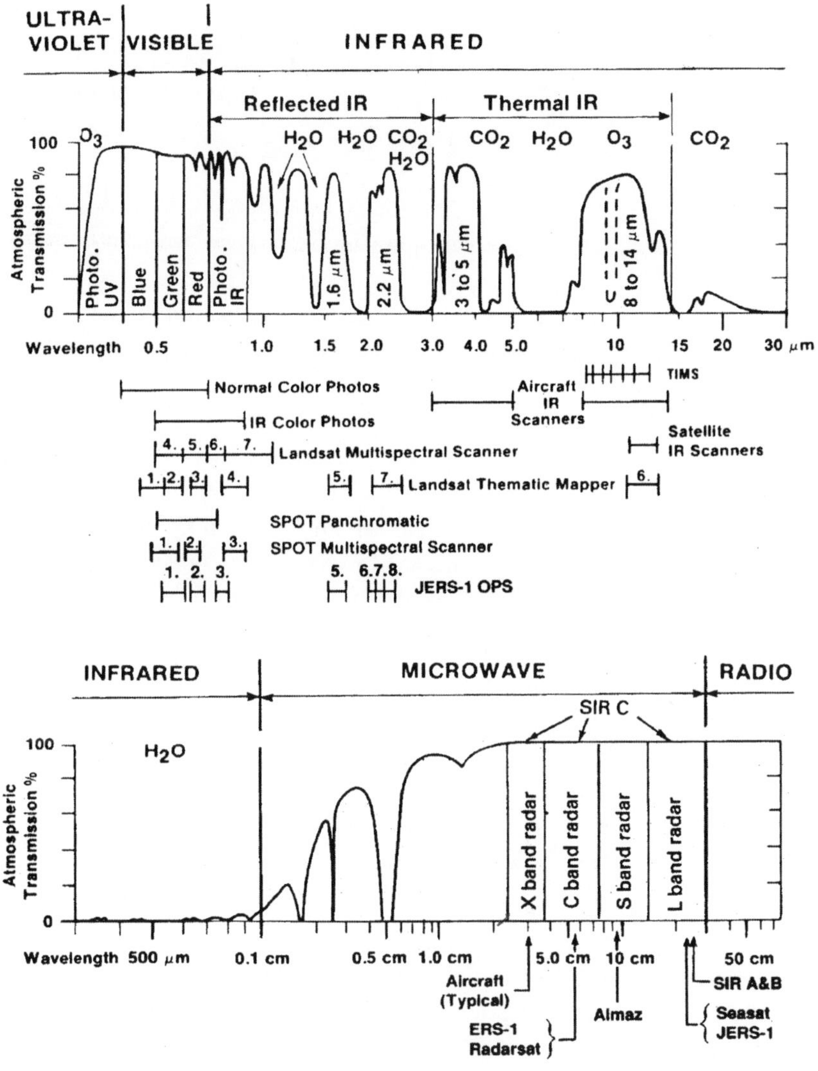

Figure 9.3: Diagram showing atmospheric transmission (windows) in visible, infrared and microwave regions and wavelength bands used in different remote sensing systems (*Source*: Sabins, 1997)

When solar radiation hits a target surface, it may be transmitted, absorbed or reflected. Many remote sensing systems operate in the wavelength region in which the reflected energy predominates. The reflectance properties of the Earth's surface features may be quantified by measuring the fraction of incident energy that is reflected. This is measured as a function of wavelength and is called spectral reflectance curve. The configuration of spectral reflectance curve gives insight into the spectral characteristics of an object based on which it is identified. The typical spectral reflectance curves of soil, vegetation and water features are given in the Fig. 9.4. The spectral reflectance curve for healthy green vegetation manifests the *peak and valley* configuration. The soil curve shows considerably less *peak and valley* variation in reflectance. The most distinctive characteristic of water is the energy absorption at near infrared wavelengths. Though these broad feature types are normally spectrally separable, the degree of separation differs in different wavelength regions. For example, water and vegetation might reflect nearly equally in visible wavelengths, yet these features are almost always separable in near infrared wavelengths. The spectral reflectance curve serves as a unique signature of the feature. In principle, a feature can be identified from its spectral reflectance signature if the sensing system has sufficient spectral resolution to distinguish its spectrum from those of the other materials. This premise provides the basis for multi-spectral remote sensing. Even within a given feature type, the proportion of reflected, absorbed and transmitted energy will vary at different wavelengths. Thus two features may be distinguishable in one spectral range and be very different in another wavelength band.

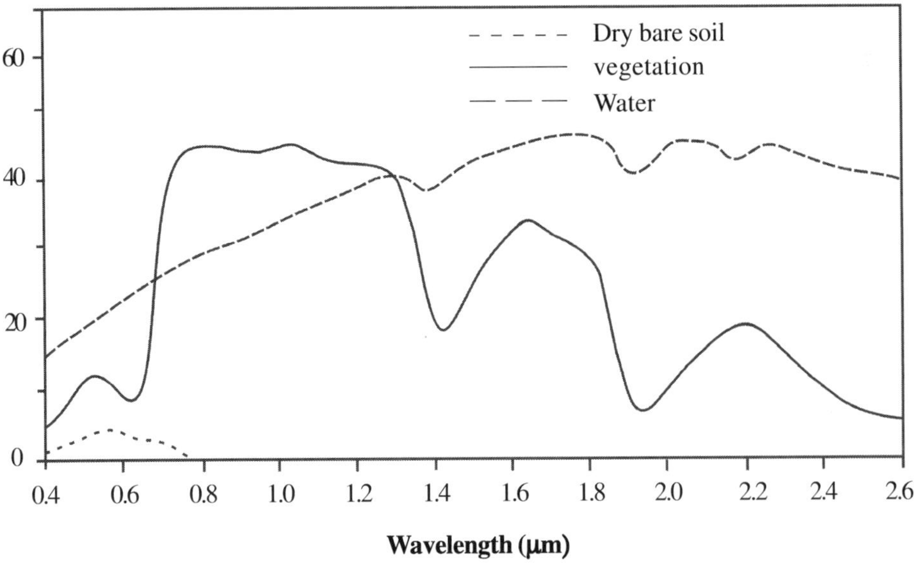

Figure 9.4: Typical spectral reflectance of soil, vegetation and water

Remote sensing images are representations of parts of the Earth's surface as seen from space. The images may be analog or digital. Aerial photographs are examples of *analog images* while satellite images acquired using electronic sensors are examples of *digital images.* A digital image comprises a two-dimensional array of individual picture elements called pixels arranged in columns and rows (Fig. 9.5). Each pixel represents an area on the Earth's surface. A pixel has an intensity value and a location address in the two-dimensional image. The intensity value represents the measured physical quantity such as the solar radiance in a given wavelength band reflected from the ground, emitted infrared radiation or backscattered radar intensity. This value is normally the average value for the whole ground area covered by the pixel.

The information acquired through remote sensing imaging is resolution dependent. Several types of resolutions are considered in remote sensing studies and those are (i) spatial resolution, (ii) spectral resolution, (iii) radiometric resolution and (iv) temporal resolution.

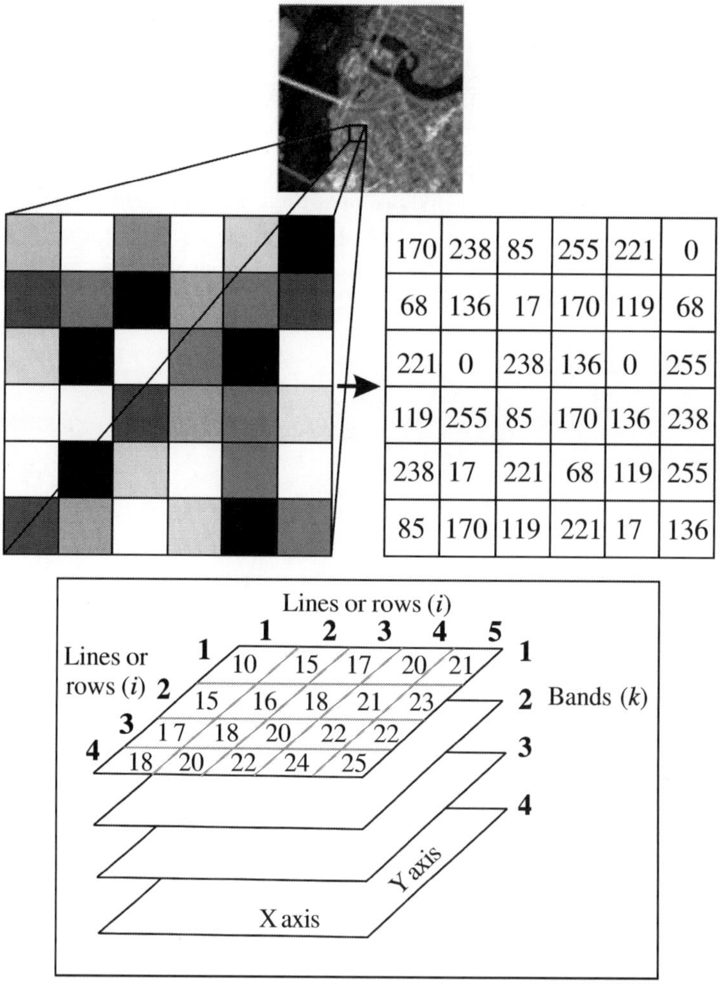

170	238	85	255	221	0
68	136	17	170	119	68
221	0	238	136	0	255
119	255	85	170	136	238
238	17	221	68	119	255
85	170	119	221	17	136

Figure 9.5: A remote sensing image showing pixels with intensity values and location address

Spatial resolution It refers to the size of the smallest object that can be resolved on the ground. In other words, it is the minimum distance between two objects that a sensor can record distinctly. In a digital image, the resolution is limited by the pixel size, i.e. the smallest resolvable object cannot be smaller than the pixel size. The intrinsic resolution of an imaging system is determined primarily by the instantaneous field of view (IFOV) of the sensor, which is a measure of the ground area viewed by a single detector element in a given instant in time. However, an IFOV value is not in all cases a true indication of the size of the smallest object that can be detected. An object insufficient contrast with respect to its background can change the overall radiance of the given pixel so that the object becomes detectable. Spatial resolution of a remote sensing system must be appropriate if one is to discern and analyze the phenomenon of interest. To move from detection to identification, the spatial resolution must improve by about three times. To pass from identification to analysis a further improvement in spatial resolution of ten or more times may be needed.

A *high resolution* image refers to one with a small resolution size. Fine details can be seen in a high resolution image. On the other hand, a *low resolution* image is one with a large resolution size, i.e., only coarse features can be observed in the image.

Spectral resolution It is determined by the band widths of the channels used in the imaging system. High spectral resolution is achieved by band widths which collectively are likely to provide more accurate spectral signature for discrete objects than by broad bandwidths. Spectral resolution varies from a single band panchromatic system, four or seven multi-spectral band system in IRS, SPOT and many satellite systems to many hyper-spectral bands of TERRA or AQUA MODIS satellite system.

Radiometric Resolution It refers to the smallest change in intensity level that can be detected by the sensing system. The radiometric resolution of a sensing system depends on the signal to noise ratio of the detector. In a digital image, the radiometric resolution is limited by the number of discrete quantization levels used to digitize the continuous intensity value. With a given spectral resolution, increasing the number of quantizing levels or improving the radiometric resolution will improve discrimination between scene objects. Interdependency between spatial, spectral and radiometric resolutions for each remote sensing system affect the various compromises and trade offs.

Temporal resolution It is an important consideration while determining the resolution characteristics of a sensor system. It is defined by the repetitive period of the sensors. This is very important to monitor any temporal dynamics of the features. For example, temporal growth profile of a crop monitored through remote sensing derived parameters helps in identifying and discriminating it.

9.4 Remote Sensing as a Data Source

As remote sensing is the non-invasive and non-destructive gathering of information about the features, it has become a valuable source of input for GIS databases because remote sensing data (in particular satelliteborne) provides synoptic viewing, data comparability, repeat coverage, the capability for historical record, as well as the provision of data in the non-visible part of the spectrum (Townshend 1981, Cooke and Doornkamp 1990).

Again satellite systems such as SPOT, NOAA-AVHRR and IRS-WiFS acquire data for large areas in a short time period, there by providing essentially uniform coverage with respect to data and level of detail. Such data are already in digital form and are provided more or less standard formats. These data products are available for almost all the Earth's land areas and are inexpensive relative to alternate sources. Although imageries are not planimetrically correct, pre-processing can often bring data to acceptable levels of geometric accuracy with only modest effort.

There are five main sources of information that can be exploited by remote sensing systems. These relate to variations in the recorded signal as a function of (i) wavelength (spectral), (ii) angle (directional), (iii) wave polarization, (iv) location (spatial), and (v) time (temporal). These are discussed below.

There are two main ways in which the relationship between surface properties and spectral response can be exploited. At one level, the aim is to discriminate different type of surface materials. In this case, the objective is to identify those wavelengths at which the contrast between their reflectance, emittance, or scattering characteristics is maximized. Since all surface materials can be differentiated at a given wavelength, it is common to record data in several parts of the electromagnetic spectrum (i.e. multi-spectral remote sensing). A subsequent aim may be to identify the nature of the surface material, i.e. to assign each a label from a set of pre-defined classes, typically expressed in terms of land cover. The second major use of multi-spectral data is to estimate values for selected properties of the observed surface materials such as to derive information on the biomass above the ground and Leaf Area Index (LAI) etc., of vegetation. This is commonly based on linear combinations of data recorded in two or more spectral bands preferably in red and near infra-red wavelengths which is otherwise referred as vegetation indices (Myneni and Williams, 1994). The enduring attraction of vegetation indices despite their well-known limitations lies in their conceptual and computational simplicity. For example, Normalized Difference Vegetation Index (NDVI) has been received enduring popularity for mapping and monitoring vegetation at regional and global scales (http://free.vgt.vito.be). Attempts have been made to develop physically based models to account for the optical properties of the leaves for their biophysical properties (Jacquemoud and Baret, 1990). These models are less data dependent and site specific and can be inverted to derive biophysical properties of the vegetation (Goel, 1989, Asrar 1989).

The detected reflectance of most surface features varies as a function of the angles at which they are illuminated by the Sun and received by the sensor. The angular distribution of the reflected radiation is described by the bi-directional reflectance distribution function (BRDF). Various mathematical models have been developed and used with BRDF for estimation of biophysical properties of vegetation and soil through model inversion techniques (Goel 1989, Barnsley et al., 1994).

Electromagnetic radiation considered in waveform has two fields (electric and magnetic) that are perpendicular to each other and also to the direction of propagation (Rees, 1990). The orientation of these two fields is known as the wave polarization of the radiation. The polarization gets affected as a result of scattering and reflection within the atmosphere and at the Earth surface. The majority of the studies in this area have employed data from the microwave region of the electromagnetic spectrum. For example, polarimetric radar data have been used to distinguish different stand in coniferous forest (Grandi et al, 1994) and to their biophysical characteristics (Baker et al., 1994).

The amount of radiation reflected, emitted or scattered from the Earth surface varies spatially in response to changes in the nature (type and properties) of the materials. These variations may be continuous, discrete, linear or localized depending on the controlling environmental processes (Davis and Simonett, 1991). They may be manifest at a variety of spatial scales (Townshend and Justice, 1990). The relation between surface type, surface properties and spatial variability in land leaving radiance has been exploited using measures of

(a) Texture — the statistical variability of the detected signal, typically based on the grey level co-occurrence matrix, measured at the level of individual pixels (Richards, 1993)

(b) Pattern — including the size and shape of discrete spatial entities (regions), typically land cover parcels, identified within the scene as well as the spatial relations between them. (LeGro 1990, Lam, 1990).

(c) Context — referring to the structural and semantic relations between discrete spatial entities identified within the scene (Barr and Barnsley, 1997).

Temporal variation in reflectance properties may be related to the diurnal or/and seasonal (e.g,. phenology) effects, episodic events (e.g., rainfall or fire), anthropogenic factors (e.g., deforestation or long term climate change).

As one possible source of GIS input data, remote sensing is useful in four capacities i.e. (i) as a display backdrop, (ii) for generation of thematic maps, (iii) for derivation of input variables for models and, (iv) as a real time link.

9.5 Integration of Remote Sensing and GIS

The relationship between remote sensing and GIS has received considerable attention in the literature and indeed remains the subject of continuing this discussion (Hinton 1996; Wilkinson, 1996). Much of this discussion revolves around the scientific and technical issues relating to integration of the two technologies (Ehlers et al., 1989, 1991), so that the remote sensed images can be used both as a source of spatial data within GIS and to exploit the functionality of GIS in processing remotely sensed data.

Increasingly remotely sensed information is finding a market in real application. Generally, the use of remote sensing achieves greatest success when the data are combined with other environmental data. Such spatially referenced environmental data are stored in a variety of formats and integrating data sets which are in different formats has until recently been a stumbling block in most analysis systems. In the late 1980's, integration was still seen as a research area, but gradually increased levels of integration were achieved by data conversion or simply at the level of raster or vector backdrop.

Increasingly the term integrated is attached to research papers describing the use of remote sensing and GIS. Many of the such integrated approaches have involved separate image processing systems with classified image results being either of the two such as (a) converted to vector format and imported into a vector GIS (b) transferred in the form of a classified coded image to a GIS capable of handling raster data for integration with vector map (Pathan et al, 1993), such results were obtained because the required—fully integrated approach was not possible using the systems available (Janssen et al, 1990). All users of environmental data should expect to make routine use of remote sensed data and to integrate them seamlessly with other data sets. A fully integrated system should involve a single software unit with combined processing (Ehlers et al., 1989). In other words, each object (point, line and polygon) in the vector data set must know which pixel in an image falls within / beneath it. Conversely each pixel should know which vector object it falls within (Fig. 9.6).

Figure 9.6: Vector layer of village polygons overlaid over the classified image to retrieve village level statistics of different crops grown

9.5.1 GIS IN IMAGE PROCESSING

The pixel is the vehicle for integrating data between GIS and remote sensing. In the terminology earlier used, the pixel is equivalent to the raster cell. Yet whereas remotely sensed imagery always conforms a continuous field model, geo-information may refer to the phenomena according to geographical data models. Hence the extraction of geo-information from remotely sensed imagery may involve transformation of the geographical data model. Ground resolution of pixel is very important in comparison to the spatial variability of the geographical phenomena of interest. High resolution corresponds to the case in which geographical objects are larger than the ground resolution cells whereas low resolution refers to the opposite case. Low resolution leads to the mixed pixel problem where it is not possible to extract information about individual objects from the image. However, using continuous field geographical model, it may still be possible to extract the relevant information about fraction of the feature or mixtures of object classes from the imagery.

Image processing and information extraction from imagery may be enhanced using layers with existing geographical data. A fundamental requirement for integrated processing of remotely sensed data and GIS data is that they may be spatially registered. The registration itself involves the actual geometric transformation of the input data set to the geometry of a reference data set. In case the reference data set is a geographic or geodetic coordinate system, the transformation is referred to as a rectification. GIS stored information can play an important role in the geometric transformation and radiometric correction of the remotely sensed imagery. Some examples are

(a) Production of ortho-imagery by means of differential restitution requires elevation data. A GIS may provide these data in the form of a digital elevation model (DEM) (Baltsavias, 1996); The coordinates of ground control points required for rectification of remotely sensed imagery can be extracted from a GIS storing infrastructure information.

(c) Slope and aspect data derived from a DEM can be used to correct for the effect of topography on the radiometric response of terrain features. (Congalton, 1991).

GIS stored information is also used in image classification in form of selecting training areas. Overlaying vector layer, for example, land cover features collected through Global Positioning System (GPS) can be used to compile a training set for the classifier (Fig. 9.7). Ideally, integration of GIS and remote sensing would allow having instantaneous access to information that supports taking a decision on which areas to include in the final training set (Sahoo, 2004). Classification of remotely sensed imagery is essentially a matter of deciding about a pixel's class label on the basis of incomplete knowledge. Incorporation of attributes derived from ancillary data sources may considerably contribute to the accuracy of the classification. This is true if there is spectral overlap amongst classes. Data concerning geology, topography, soils, vegetation or some other features may improve discrimination of these classes. There are several ways to make use of ancillary data, which can be categorized into pre- and post classification techniques and classifier operations according to the ancillary data's function in the classification process.

Use of ancillary data preceding classification involves a stratification of the image according to some criterion or rules. The purpose of the stratification is to reduce the variation within strata and thus improve class distinction. Post classification technique involves some sort of rule based on re-labeling of image classification which results in improving the accuracy. Strahler (1986) showed how ancillary knowledge could be effectively incorporated into image classification through the use of modified prior probabilities in maximum likelihood classification.

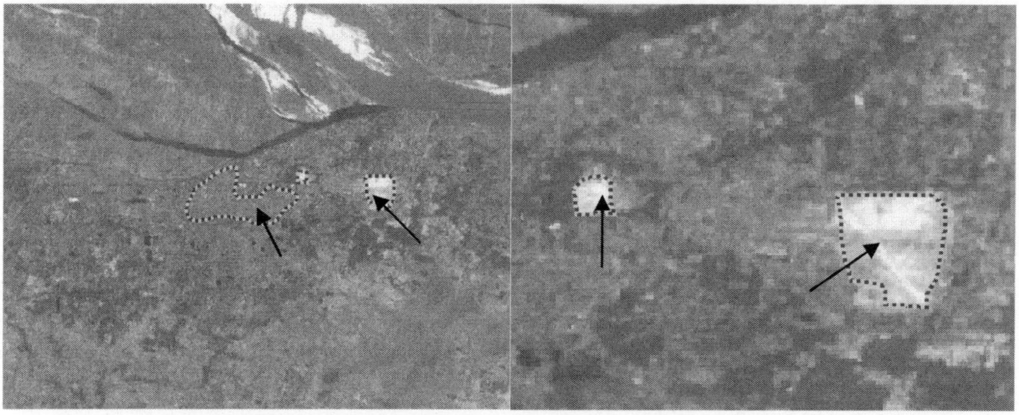

Figure 9.7: Vector layers as training sites overlaid over the imagery for classification

In an experiment involving classification of natural vegetation, the use of prior probabilities contingent to elevation and aspect classes resulted in a considerably improved classification accuracy over that obtained with spectral data alone (Strahler, 1986). Janssen and Middelkoop (1992) used a similar approach to incorporate historical land cover data stored in a GIS and knowledge on crop rotation schemes in the classification procedure. They also reported improved classification accuracy when prior probabilities were used. Developed an extension to maximum likelihood classification that uses classification results to iteratively adjust spatially distributed prior probabilities. Their procedure does not require the priors to be known at the outset, but estimates them from the image to be classified.

Hinton (1999) developed a methodology to perform a classification of radar image data of a forest by combining the image data with vector and tabular information. The calculation of image mean within each polygon, for use in the classification removes the effects of speckle in the image and the noise it introduces into the classified image. This per-polygon classification removes the need for image filtering operations, which would blur the boundaries, raster editing and image classification. The ability to customize the system to specific requirements is also a significant advantage and makes the system as flexible as possible. In a fully integrated GIS, it is possible to implement this classification methodology for a selected set of polygon only either selected manually by the user or as a result of some GIS query.

9.5.2 Geographic information extraction from remote sensing

Image classification results in a field model of the terrain in the sense that posterior probabilities and class labels are treated as functions that take a value at any position in a plane. This model does not match the often pursued objective of representing geographical objects such as parcels and water bodies, in a GIS database. To achieve the latter, it is necessary to subdivide an image into different segments that correspond to meaningful objects in the terrain. These objects can then be subjected to further analysis or interpretation (Wahl, 1987). Sometimes, object geometry can be derived from the image itself by a technique called image segmentation.

The extraction of exact objects from remotely sensed imagery always requires high resolution imagery in the sense that terrain objects are represented by several pixels. Otherwise, if the objects are smaller than the pixel, it will be necessary to resort to a continuous field geographical model (Strahler et al, 1986). By doing so it may be possible to extract information about mixtures of object classes. A mixed pixel may represent a ground resolution cell that comprises more than one discrete object class e.g. NOAA AVHRR imagery (pixel size: 1.1 km) or SPOT vegetation data (pixel size: 1km). Conventional classification fails to extract the information from mixed pixel. It would be more appropriate to use spectral unmixing technique for this purpose. The success of spectral unmixing strongly depends on the separability of the component classes' spectral signatures (Sahoo, 2004).

A mixed pixel may also represent a ground resolution cell in which two or more classes of mixed composition gradually inter-grade. This representation corresponds with the fuzzy object geographical model, which recognizes that objects may not be exactly definable (Foody, 1992). Their components belong to partially overlapping classes having diffuse boundaries in attribute space. A commonly applied method to derive membership to fuzzy classes is fuzzy c-means classification (Bezdek et al., 1984). This technique has been frequently used in soil science (Burrough et al., 1997).

As the sizes of the objects become increasingly small compared to the size of the resolution cell, the applicability of the continuous field geographical model becomes increasingly evident. Instead of observing individual objects it becomes more appropriate to consider concentrations, which vary continuously in space. For example, it makes no sense to study individual suspended sediment particles in coastal waters using remote sensing, but it makes a great sense to use remote sensing to estimate suspended sediment concentrations. Used least squares estimation to derive a regression equation relating sparsely sampled suspended sediment concentration to radiometrically and geometrically corrected Landsat MSS data. Once a suitable regression equation is found it can be used to estimate the response variable from the imagery, for which area coverage is complete. But serious drawback of using regression for spatial estimation from remotely sensing imagery is that it takes no account of the autocorrelation in a scene.

9.5.3 Problems with Integrating Spatial Data Sets

The effective integration of multiple data sets is a complex process that can be hampered by numerous problems. These problems have been well documented in the general GIS literature (see Ehlers 1992). It is only very recently that some workers, for example Gong et al (1996), have considered the problems of digital data synergy. The primary difficulties stem from the issues as described below.

The Spatial Resolution Problem For the successful integration of satellite and air borne imagery and GPS data sets, the researcher must consider possible constraints caused by the differences in their various spatial resolutions and the degree of compatibility between each of these data sets. This question concerns whether the different resolution data sets compliment each other. For example, can differential GPS which provides accuracy ranging from sub-meter level to 5 meters, fully synergies with 20-30 meter mid-resolution satellite imagery, and 1-10 meters aerial photographs (depending upon the resolving distance selected) to provide a coherent data set. The development of integrated systems in the future must consider the wide range of new sources of data becoming available. The use of high spatial and spectral resolution remotely sensed data is becoming common and facilities must be provided to derive maximum benefits from them.

Data Set Errors Compounded by Integration Errors are common to all spatial digital data sets. These errors are produced during the stages of "data collection, manipulation, interpretation, and presentation," such as image capture for remote sensing data and the digitizing of data for input into a GIS framework (Gong et al, 1996). Such error sources can be managed through the continual process of data quality assessment, which involves the identification of error sources and the provision of error estimates (Brusegard and Menger, 1989). Once the presence and cause of data errors have been recognized, the value of the spatial data sets can be improved through full understanding of the nature of the data set (Brusegard and Menger, 1989).

Raster-Vector Data Set Integration The implication of the raster/vector data representation of spatial data is an on-going issue of relevance to many GIS based investigations. Ehlers (1992) suggests that a fully integrated single system is the only effective way of resolving the different representations of space as given currently by raster and vector based data sets.

Matching Data Sets over Time Successful synergy of different data sets requires that data capture occurs either at precisely the same time or at least within the same season. The lack of a good temporal match frequently means that the data sets cannot be combined, or that the resultant synergised data is degraded and of insufficient quality for the applications which are dependent upon feature detection.

In addition, the lack of recognition or failure to document how data integration problems have been overcome makes the comparison of results between geomorphic studies problematic and in some cases impossible. Therefore, in the same way that geomorphologists have established tried-and-tested methods for the analysis and comparison of traditional field mapping techniques (Thorp 1986, Ballantyne 1989), there is a need for a similar approach to develop a methodology or strategy for the integrated use of multi-source digital data sets for feature detection and mapping in mountain environments. This re-assessment leads on from the work of Carver et al. (1995), which argued the need for a more effective methodology for data integration white using GIS and GPS as a field technology. Furthermore, the formalization of such a methodology will permit common data set errors to be established.

9.6 Software Options

Image processing systems are being equipped with GIS capabilities while GIS packages are coming with image processing functionalities. Many systems support both the vector and the raster data structures. Remote sensing imagery as a stack of raster layers that may be co-registered with vector layers having information like road network, cropped lands etc. Vector layers may be easily converted to the raster structure to enable integrated analysis. The software commonly available for image processing are ERDAS Imagine, ENVI, IDIRISI etc. and for GIS analysis are ARC/INFO, ARC VIEW, SPANS, MAP INFO, having modules for raster and vector analysis and integrating them for information extraction.

Much software has high level scripting languages that the allow complex data processing tasks to be accomplished with relative ease. Examples are Arc Macro Language (AML) of ARC/INFO, Spatial Modeler Language of ERDAS Imagine, and Interactive Data Language (IDL) in ENVI. Although these languages provide suitable tools for building special purpose applications, they imply a duplication of programming efforts if the required functionality is available in another system. Also, interpreted scripts generally execute slower than compiled software modules.

Sometimes it is preferred to use different programs in complex spatial analysis. Until recently, the only way to let these programs communicate was to transfer disk-stored data sets between them. This often required storage of multiple copies of the same data in different formats, easily causing data integrity problems. Recent advances in software engineering such as those supporting cooperating processes, and the move to open system architectures (open GIS), provide major enhancements for integrating multiple systems (Woodsford, 1996). By setting standards for the database parts of GIS, software from various companies can be used to display and query data that may even reside on remote computers (Burrough and McDonnell, 1998). Object enabling system software enables separate applications to be smoothly integrated with other applications or packaged software. Eventually, these technologies will enable specially tailored GIS solutions to be built out of custom components and software modules from a variety of vendors (Kemp and Vckovski, 1998).

Despite numbers of mutual benefits from integration of remote sensing and GIS, actual progress towards the goal of full integration is surprisingly slow. This is due to very considerable technical challenge of accessing, manipulating, and visualizing vector, raster and tabular data simultaneously. It may not be the sole barrier to achieve the full integration. It may be due to drawing attention away from the issue of integration to various environmental simulation modeling for global environmental monitoring and mapping which are closely allied to the concerns and use of GIS. Similarly recent developments in GIS functionality within standard software for wider access to GIS through network and web platforms and multimedia technology in GIS have consumed much of research and development effort in the field of GIS.

Major developments are now taking place in the integration of remote sensing data with GIS. Techniques such as the use of knowledge based systems and artificial neural networks appear to offer much potential for the extraction of improved geographical information from current remotely sensed satellite imagery. Whilst the use of such techniques offers much promise in environmental monitoring and management, a number of fundamental problems still remain such as choice of appropriate data structures, and procedures for handling error and uncertainty. Over the next ten years the complexity of remotely sensed datasets will grow significantly through the use of multi-sensor, hyperspectral, and multi-view angle approaches besides use of time series. Current GIS techniques and technology are not appropriate to handle the much increased dimensionality of such datasets and new developments are needed both in visualization tools, and in spatial and temporal analysis.

However, for two reasons, integration of remote sensing and GIS will receive fresh impetus in near future. The first is the increasing availability of data from very high spatial resolution from government and commercial satellites. This will produce data appropriate to large scale GIS applications where it would compete directly with the traditional aerial photography market. The second is the high resolution satellite data products would need new data processing algorithms to extract the maximum amount of information. There is considerable overlap between these techniques and those used in mainstream GIS at least in terms of their potential for spatial analysis and this would bring better integration of remote sensing and GIS in future.

CHAPTER 10

ISSUES IN GIS

"Concern for man himself and his fate must form the chief interest for all technical endeavors. Never forget this in the midst of your diagrams and equations."

Albert Einstein

10.1 INTERNET AND GIS

In early days of GIS, users could transfer data among themselves using physical media - nine-track reel tapes or other physical storage devices. The transfer of the data was a cumbersome process, thereby increasing costs and impeding the development of GIS in many ways. Basically, there were UNIX and VAX machines running on mainframe and minicomputers. The PC GIS users could not use these storage devices and the data were mainly stored in floppy disks. These small capacity disks can store a very small volume of data. Thus both the storage and transfer of the data was very much limited.

The last decade has witnessed some radical improvements in developing storage devices with very high capacity, which has been facilitating the storage and transfer of data with increasing efficiency. But we must agree that it is the *Internet,* which transformed the way in which spatial data is transferred. The size of the files has increased, and also the speed of the Internet, allowing a fast transfer of data. In the last few years there has been an increase in the use of the spatial data with decreasing cost of software and hardware components and therefore the access to the computers has increased. These have resulted in more public awareness and attention. Of late, people have begun to realize the importance of Internet as a medium for rapid and efficient exchange of information both at home and at workplace. Private and public organizations including Universities and Institutions started developing their own web sites/web pages and post them onto the Internet. Moreover, they also set up not accessible by anyone outside the organization, but are incredibly useful for sharing data within the organization. Internet has also emerged as a tool to communicate, advertise and even to shop, thus reaching to a wider group of people. It also facilitates introducing web and email sign-up offers, free email and browsing services (e.g. Yahoo, Google, Hotmail), diskettes or CD-ROMs e.g. CompuServe and AOL containing Internet browser software (Netscape Navigator, MS Internet Explorer, Opera etc.) as well as software tools (HTML editors, image processing). It has also reduced the error and problems caused by the update of multiple versions of the same spatial data file, say, by the assessors department and the public works department of a country. GIS Intranets allow for a map library where data is checked out, updated, and checked back in to one master database.

The Internet has also transformed the way that spatial data is served from producers to users, beginning in the 1980s with file transfer protocol (FTP) sites. These sites allowed users to break away from handling physical media. Next, HTML pages in the 1990s on the World Wide Web (www) allowed users more flexibility and ease of use. Then, continuing in the late 1990s to till date, FTP sites, web pages, and Internet Map Services (IMS) sites brought additional capability into the hands of GIS users. IMS allows the data producer to host spatial data holdings of an organization online so that others can download them, browse them, and even ask questions (queries) of the data with an ordinary web browser. They often can also download from these sites. The software allows users to integrate their own data on their own computers with data on the Internet, for display, query, and analysis. Today, a data user can use data on IMS sites without downloading while the data is still online. For example, with ArcGIS, a theme may be added while on the Internet, such as real-time weather data. Many good examples of IMS sites are:

1. The Sample IMS gallery at ESRI: *http://maps.esri.com*
2. The near-real time wildfire site: *http://www.geomac.gov*
3. The National Atlas Project: *http://nationalatlas.gov*

These sites can be looked into for more information on the topic discussed above.

Remote sensing (RS) and GIS-related subjccts are well represented on the web for several reasons. Much RS/GIS data exist in computer-compatible format, and many users are familiar and comfortable with working in a network, computer-based environment. The subject matter is image rich, which lends itself to what the web does best—delivering text and graphic imagery. Internet also has the ability to move large files quickly (FTP)—a great advantage for transferring GIS databases from one location to another. And lastly, the web offers user interaction, so that a distant user can access, manipulate, and display geographic databases from a GIS server computer. Two examples of web interactive GIS are:

- Montana Natural Resources Information System—Interactive Map Builder
- Kansas Data Access and Support Center—DASC

10.2 Geographical Objects with Indeterminate Boundaries

Any geographical phenomena requires two descriptors—*what is* and *where is*, in order to represent the real world. The former may be described as well-established objects or entities, whereas for the latter, the referencing may be defined in terms of a geometrically exact or a relative location. As we have seen, descriptors are described in many ways depending upon the individual's perception and means to describe them. But in all the cases, it is necessary to divide the world (or the subset of the world) either into crisp *entities* (landform, soil unit, administrative boundaries) or *continuous fields* (atmospheric temperature, groundwater levels, soil organic carbon content). In the first approach, defining the entity is the first step; listing the attributes and defining locations comes second. In the continuous fields approach, the attribute and its spatial variation comes first.

The philosophical basis behind the division of geographical space/ phenomenon into smaller well-defined objects is that of reductionist approach (*reducttio ad absurdum)*. This approach, deeply embedded in western scientific thought, calls for the division of a whole problem into a set of smaller problems and the ultimate solution would be the aggregation of all sub-solutions. The axioms behind this approach are based on Aristotle's 'laws of thought', namely—(a) law of identity (everything is what it is—a tree is a tree), (b) law of non-contradiction (something and its negation cannot be both a tree and not a tree) and (c) principle of the excluded middle (every statement is true and false—a tree is evergreen or it is 'not').

The principle of excluded middle ensures that any statement must be either *true* or *false*—an assumption lies at the heart of most of the Mathematics and Computer Science (Burrough, 1997). Till date, their principle is always followed in developing tools for computation and retrieval of data (e.g. SQL). Hence there was no concept of partial or ideas of class overlap or partial membership of a set.

Of late, people started in realizing that this simple binary logic limits the expression in the way we see the world. It is more in geographical phenomenon that most of the times things overlap in spatial and temporal scales. There is geographical phenomenon that possesses attributes—the geographical boundaries often do not match with the boundaries in attribute space. These phenomena also vary continuously with a different mixture of attribute values being observed at each separate sample point. Let us visualize the problems associated with binary logic of data retrieval.

Say, we try to find out the erosion hazards of a piece of land. If the criteria or the parameters are defined as slope of land, soil texture, vegetation cover and the presence of hard pan, then severity classes are determined based on a combination of these parameters. Here we apply Boolean operators AND, OR, NOT, and XOR. Thus, a severe erosion class may be defined as:

Slope ≥ 10% AND (soil texture = sand OR presence of hard pan) AND vegetation cover £ 25%

Now, the question arises —What would be the result if the slope is 9% or the vegetation cover is 27%? Definitely as per logic and rule, erosion is not the severe. But is it enough? Certainly not. Here we are excluding the regions, which fall marginally to be included in severity class, thereby underrating the problem of soil erosion as a whole? If we think of the other side, how precisely are we able to measure these parameters in the field? It is indeed difficult to measure land slope to a precision of 1% and so the textural classes. It is far more difficult to define the vegetation cover. Even if theoretically it is defined, it is impossible to measure the same in the field. These all force us to go for certain approximation leading up to interpret the results often wrongly.

Taking this problem to another dimension, the physical or actual boundaries of the classes (when represented in a map) is very 'indeterminate'. More so, when we consider vector GIS as the right tool for storing and manipulating soil information. This approximation in defining sharp boundaries leads to well-known difficulties in GIS processing (i.e., point-in-polygon and overlay operations, Burrough, 1986). Furthermore, some geostatistical analyses have shown (Voltz and Webster, 1990) that taking into account the discontinuities in the soil cover strongly improves the interpolation of soil properties. Again, many data (in soil study) collected during field survey are often described as *less fertile* or *poorly drained* or *highly calcareous* or *less susceptible to erosion,* all seemingly vague terms to interpret with.

10.2.1 Fuzzy sets and fuzzy objects

Lotfi A. Zadeh, the professor at University of California at Berkeley was the first to propose a theory of fuzzy sets and an associated logic, namely *fuzzy logic* to deal with inexact concepts in a definable way (Zadeh, 1965). Essentially, a fuzzy set is a set whose members have membership degree between 0 and 1, as opposed to classical sets where each element must have either 0 or 1 as its degree of membership. As classical logic is based on classical set theory, fuzzy logic is based on fuzzy set theory. The name fuzzy has nothing to do with unstructured, wooly or haziness; else it is a type of imprecision characterizing classes that for various reasons cannot be precisely defined. "He is about 20 years old" is a perfect fuzzy statement, as the age, when told in terms of years, does not have precision (temporary, how many months/days/hours/minute!). Fuzzy sets whenever used deal with ambiguity, vagueness and ambivalence in mathematical or conceptual models of empirical phenomena.

10.2.2 Crisp sets and fuzzy sets

A crisp or Boolean set is a set in which all members match with the class concept and the class boundaries are sharp. The degree to which individual observation z is a number of the set is expressed by the 'membership function', MF^B. This function for crisp sets can take the value of 0 or 1.

$$MF^B(z) = 1 \text{ if } a1 \text{ £ } z \text{ £ } a2$$
$$MF^B(z) = 0 \text{ if } a1 > z \text{ or } a2 < z$$

Where $a1$ and $a2$ are the exact boundaries of the set.

Unlike crisp sets, which allow binary membership functions (an individual is either a member or not a member of the set), fuzzy sets, admit the possibility of 'partial membership function'. A fuzzy set can be written as

$$A = (z, MF^{B}(z)) \text{ for all } z \hat{I} Z$$

Where the membership function MF^{B} (z) is known as the grade of membership of *z* in *A* with values ranging 0 to 1 (in contrast of only two values - 0 or 1). The membership function tells to what degree an individual *z* is a member of set *A*. Value 1 represents full membership and a zero value indicates no membership.

Membership functions for fuzzy sets are chosen on any of the two approaches: (1) based on an *a priori* membership function with which individuals can be assigned a membership grade (similar to the expert knowledge used in Boolean logic) or (2) analogous to cluster analysis and numerical taxonomy where the value of the membership function of the classifier is used.

Logical operations of data in fuzzy sets are just as Boolean algebra (i.e., AND/OR/NOT/XOR and other operations used in Boolean sets (Kandel, 1986; Kauffman, 1975; Klir and Folger, 1988). Standard query language in relational database management systems has also been modified to accept continuous operations (Kollods and Voliotis, 1991). For example, the AND operation in binary logic is replaced by 'MIN' operation; the OR by a MAX operation. In the example of erosion hazardnesss, a severe erosion class may be defined as (in fuzzy logic).

Result = MIN [slope, MAX (texture, hard pan), vegetation cover]

All these parameters must take a value between (and including) 0 (almost no threat of erosion) and 1 value (sufficiently high to be justified as erosion prone).

There is a growing literature illustrating the practical advantages of the fuzzy approach for geographical data processing using GIS. There seems, however, to be a relatively poor understanding of the distinctions between Boolean and fuzzy logic, the latter often equated with probability. Many authors failed to provide a clear distinction, and those who set out the differences between these two, have not received due attention they deserved. Probability cannot be appropriate to define geographic objects with indeterminate boundaries since the probability is a Boolean phenomenon or region does not have undermined indeterminate boundaries in space. It has definite boundaries, which may be hard to measure with precision.

10.3 Legal Issues

By definition, the word *legal* means to established by or founded upon by law or official or accepted rules. The *law* is a set of rules or principles dealing with a specific area of a legal system; thus the legal system has to be in accordance with the forms and usages of law. This has also got the constitutional approval so that any legal act, if not constitutional, is not unlawful. Laws must be meant for resolving disputes, ensuring order in the society, maintaining efficiency, harmony and balance in the functioning of government or civil bodies, protecting the weak against abuse of power and above all, providing equal opportunity to enjoy the minimum decencies of life.

Once we refer to GIS, we need to talk in terms of information. The very first point to consider is whether information is regarded as a public good. In strict economic sense, a public good must be freely available and also its use should not to be restricted or curtained at any point of time. But as on today, most of the information and GIS product are not free goods, neither they are available in public domains (except some data producted by Government Agencies). As there are more or less restrictions in use of GIS data/ product, we need to consider them as priced commodity. But whether information would be considered as a commodity is debatable. For a traded commodity, like automobiles or garments, or even food produce, the value is usually proportional to their respective quantities. For information, this is rather difficult. Though many times the quantities of information do concern (as related to the extent of geographical area covered); but it is the quality that really matters. Lack of good quality information leads us to wrong decision making and with lots of uncertainty. On the other side, like any commodity, information can also be purchased, is movable/transferable and can be value added. At the same time, both the amount and the value to be associated in getting the information are difficult to measure.

Our next concern is related to property issues. The concept of *property* (or *ownership*) has no universally accepted definition. It is a fundamental concept that has a great value in day-to-day public and private discourse, yet the usage of the term varies widely between fields of laws, economics, sociology etc.

To a common man, *property* is understood as somebodys own thing. In a broader sense it includes the complex relationships between an *owner* and their property and also the rights and obligations attached—the rights as defined by law. Apart from individual ownership including government, corporations, companies etc., there are *commons* (belong to a defined community) and *public* (where access is unlimited). The World Trade Organization (WTO) agreement on Trade Related Aspects of Intellectual Property Rights (TRIPs) is an international treaty which sets down minimum standards for most forms of intellectual property regulation within all member countries of the WTO.

In relation to property rights, role of information as a commodity/ service is confusing. It is difficult to put information in context of property issues, as it is difficult to determine – *who owns information*' and *how long*? Unless these are well-defined, property rights regime cannot be applied to information. Again, without a proper protection system, people may not find to produce and deliver information as viable and profitable. The existing patent system cannot be applied as such to information, as the system can never prevent misuse of information—thus making difficult to assign to information paradigm.

Another problem with information is that often we are not aware of the forms and accuracy of data and we still take decisions. There is no contact liability (according to legal laws) between the supplier of data and the user. Moreover there is no proper guidance and standardization of design and procedure for data capture, data conversion and software use—thus quality of processed data varies widely. As discussed above, the ownership issues are not well-defined, and this is more so when the data is converted from analogue to digital form.

Though any data is subject to reservation to personal privacy (see Section 10.5) and therefore restricted use, laws should enforce the availability of data to the maximum extent. Protection by laws against inappropriate and misuse of personal data should be guaranteed and more so, once the data is available in digital form. The complicacies of open and commercial access of data with privacy and copyright issues make the legal system related to GIS very difficult to implement.

10.4 Errors in GIS

Before getting into this discussion let us clarify, at the very beginning, the difference between *accuracy* and *precision*, two terms quite often used to designate errors. Accuracy is the degree of trueness, i.e., how the data/information is close to the exact or true value. Precision, on the contrary, refers to the level of measurement i.e., how exact is the description. Accuracy tells us the quality of the data and information contained in a map. However, the level of accuracy varies and so the form (spatial/ non-spatial accuracy, conceptual accuracy and so on). Undoubtedly, a precise data provides information of a feature in more details. But similar to accuracy, the level of precision required for particular applications also varies. A very precise measurement may be required to obtain the quality difference of grains between two cereal species, but the same may be impractical in getting the yield difference. Data quality in GIS refers to the relative accuracy and precision of data used. High precision does not imply high accuracy and vice-versa. Highly precise and highly accurate data come with a high price tag—thus a trade off is always necessary to plan the quality of data to be used for a particular GIS project.

Errors in GIS refer to the use of imprecise and inaccurate data as well as the procedures of analysis (including data capture, data conversion and data retrieval). The major role of GIS is to relate many types of data by location, facilitating a very speedy and efficient data analysis. This strength of GIS to collate data linked to a single location/ reference is also the major source of error in GIS as this *positional* error is considered to be of great concern today. Every time a new dataset is imported or existing datasets updated, the errors are also innate. However, apart from Positional and other errors have been discussed below.

Positional errors Positional accuracy is a measurement of the variance of map features and the true position of the attribute (Antenucci et al., 1991). According to USGS, the horizontal positional accuracy of a map should be so as 90 per cent of all measurable points must be within 1/30th of an inch for maps at a scale of 1:20,000 or larger, and 1/20th of an inch for maps at scales smaller than 1:20,000. The similar is the vertical accuracy, which together with horizontal accuracy defines the mapping standards. Accuracy and precision depends on the scale at which a map is created. Thus, a point or a line on a map cannot be at its absolute location but have its *probable* location within a certain area.

There is another very typical error, which is often outthought. The user may have a tendency to look into details using *zoom in* or *zoom out* facilities while working with a virtual map. In this process he assumes that the scale is also changing and so the accuracy and precision are improving, which is not so. The accuracy and precision at which a map is made is inherent and tied to the original map scale. These do not change. The apparent improvement of accuracy and precision is false, which lead to a very potential error.

Errors related to attributes Just like positional information, non-spatial or attribute data linked to position may also lack accuracies or precisions. These inaccuracies may be very much human error while in entering datasheets. However, these mistakes are seldom noticed, but give wrong output decision. Sometimes, the precision at which the attribute information should be included in a database is absent. These incomplete attribute data provides partial information making it difficult to come to some inferences. For example, without proper information of climate and soil type or variety, any conclusion relating yield attributes of crops to the agronomic practices would be illogical.

Conceptual errors As explained in Chapter 2 and 3, we try to model through GIS, the Earth and its features based on our own understanding. This abstraction from reality is a set of scientific procedures, yet it varies from person to person and situation to situation. The user determines how a real world phenomenon could be classified into categories and what amount of information is necessary for the same. The inappropriate classification misleads our objectives. Thus the concept by which the real world is explained through conventional computer-literate language followed by categorization is very important. For example, where the drainage network is ordered following Strahler's classification, to calculate the flow accumulation (using GIS), caution has to be taken to have a perfect drainage network digitized. If the digitization goes wrong, or the tributaries are miscounted, the whole exercise (even if ordered perfectly) will be futile. On the contrary some studies might not require this categorization of stream order – for them the location and names of all stream and rivers are sufficient, regardless of order.

It is always difficult to understand and interpret quantity of data beyond a limit. To comprehend vast amounts of data, classification and generalization is a must. These are subject to interpolation error and are hard to be detected.

Logical errors Information stored in a database must be employed carefully to yield meaningful and effective results. There are chances that the data are used illogically. For example, the flood-prone areas cannot be mapped correctly based on single date imagery. A series of aerial photography or satellite imagery is required to monitor the flood and its variations over time. GIS will fail to interpret logically if comparisons are being made inappropriately or if data are being used incorrectly. Some rules for use can be incorporated in GIS designed as *expert systems,* but developers still need to make sure that the rules employed match the characteristics of the real-world phenomena they are modeling.

According to Burrough and McDonnel (1997), sources of errors can be categorized into (i) obvious sources of error, (ii) errors resulting from natural variations or from original measurements, and (iii) errors arising through processing. The third form of error is difficult to identify. A brief discussion of these typical sources of error follows.

Sometimes data sources may be too old and lack the standard precision that is being followed today. There would have been lot of changes, and modifications from the actual facts over time. Similar is the incomplete or partially available data on a given area. In areas of maximum cloudy days and heavy rainfall, availability of clear optical data is very difficult. Decision has to be taken in these cases, on level of generalization or whether further collection of data is required. Old and partial data are potential source of error in a GIS project.

Map scale is another factor, which restricts type, quantity, and quality of data (Star and Estes 1990). One must match the appropriate scale to the level of detail required in the project. Enlarging a small-scale map will not increase its level of accuracy or detail, as explained above.

GIS is known for its capability of spatial analyses. However, the number of observations required to perform a proper spatial analysis is another crucial factor. The sampling interval should be close enough to facilitate accurate interpolation.5. Relevance:

Data formatting like conversion of scale and projection, raster to vector format, or enhancement *etc.* are possible areas of error. These methods of transformation and subsequent processing should follow certain standards. Faulty and biased observation followed by miscalibration or use of inappropriate instrument may yield huge variations in data. Maps must be correct and free from bias. Mistakes made by faulty and non-calibrated instruments used to measure features in the field or laboratory may be difficult to detect during GIS analyses unless the errors lead to unexpected or conflicting results. Numerical errors and errors during digitization (discussed in Section 4.3.1.1) just add to more woes.

10.5 GIS in School Education

It is now clearly understood that GIS has already engulfed most of the arena in our life. By character, it is a technology and has links to most of educational disciplines existing on our Mother Earth. As discussed earlier, GIS is gradually giving rise to a new and rather transdisciplinary subject namely, geosciences, which need to be introduced in school education. However, GIS education requires good and powerful computers (both software and hardware), but not necessarily at the very beginning stages. With a clear thinking and understanding of GIS, one can develop some intuitive ideas as teaching aids. At the initial stags, on can very well use manual approaches like transparent overlay techniques (Walsh, 1998). This technique involves use of transparencies to demonstrate and facilitate hands-on training of the 'overlay' process. It may be a unique experiment, where each gets an equal opportunity to understand what is involved in GIS at both theoretical and practical levels (Green and King, 2003). This type of teaching hands initiates the process of primary education on GIS, until the computers are made available that can take care of the advanced studies. One should not think that manual techniques underrate the value of GIS as an information technology in a geography environment; it rather drives us to realize the importance of clear understanding of the process of collection, analysis and use of geographical (spatial) data for solving our daily and common problems.

Once computers are in, geographic information science and technology education can be imparted to the students through the use of basic concepts and tools like spreadsheets, database systems etc. and later on hands-on experience with simple GIS packages like Arc View, IDRISI. There is a number of GIS software available in the market, which gives a scope of choosing the best under a certain set of conditions. In our opinion, schools can definitely afford to have a full-fledged GIS system for classroom teaching with a modest investment, but advisable only after the students have acquired a firm grip of the basics of GIS.

GIS can be introduced at all levels of teaching without disturbing the current on-going curricula and systems. It need not be viewed as a new introductory subject, but a new way of analyzing the information, which gives students more power of thinking and finally leading to smarter decisions. GIS can be taught with GIS and so other subjects. Introducing GIS at multiple levels and disciplines in schools can truly bridge the gaps in exploration, integration and cooperation between students, teachers and society for a mutual benefit of all.

10.6 GIS and Decision Support

We could see in our definition of GIS (Chapter 1) how spatial decision-making takes a place there. The general structure of any decision-making process may be, according to Simon (1960) consisting of three phases – intelligence (whether the problem exists or there is an opportunity for change), design (working out the alternatives) and choice (to choose the alternative – best among them). Generally, the decision-maker follows the path intelligence®design®choice, but time and again he/she has to go against the loop to get a proper feed back to correct/modify his steps in the decision-making process. One's intelligence may not be enough to achieve the final decision and similarly, the design has to be looked back once it is found not implementing. Let's now look into what and how GIS plays a role in the decision-making.

The need for decisions are felt only when a problem exists and there are opportunities to change or improve upon. We feel for a judgment when we realize that the situation is not as it should be and there is a scope for improvement. A proper decision aims to bridge the gap between the 'desired' or 'ideal' and the 'existing'. In this phase raw data or information need to be collected, processed, analyzed to find out any 'clue' that may define the alternatives or the solutions. It is essential to choose among datasets, which are relevant and also what other datasets are required. What else than GIS-based techniques should be the best possible way to effectively store, retrieve, analyze the data and observe the output in the most desired way? In an interactive environment, GIS is perhaps the best choice to clearly identify the recordable data items, which are useful, relevant, accurate, reliable and flexible in the context of the spatial decision problems. GIS helps in more effective and effective collection and analysis of data. It is unique in its way to handle, store and manage large volume of data, which otherwise becomes a daunting task. GIS with its ability to integrate data from a wide range of sources, help in coordinating situation analysis *e.g.* cropping system information form satellite source can best be integrated with point based soil-survey data and tabular socio-economic information to have a better picture for deciding alternate land use and cropping systems. In the present day context, performing this task without a GIS possibly can not be thought of.

10.7 Privacy Issues

Geographic information systems are emerging as a very powerful system spreading in all parts of the society with tremendous influence in private, government and other business activities. Thus any application GIS is being tagged with social responsibilities. As a technology the changes that GIS is bringing, has both positive and negative sides. The positive points are discussed in earlier Chapters, with some hints on the ill sides too. A very potential negative effect that GIS is bringing in the society is the decrease in personal *privacy*. Privacy is the condition of being concealed or hidden, which gives a person the ability to control the availability of information about him- or her. Apparently geographic information systems deal with only factual information about land and resources (spatial) – hence has nothing to do with personal or individual privacy issues. But if we look more closely, the ability of a GIS to store, integrate and analyze vast volume of data with respect to its geographic location makes it one of the most powerful systems in compiling data from diverse sources on families, households and individuals - something that was a practical impossible even a few years ago (Eitenbichler, 1993). These capabilities of GIS software make GIS highly effective tools for analyzing personal information and its potential is far more invasive of personal privacy than many other information technologies.

There are two sides of the issues, each party advocating its points strongly. However, most common people are unaware of the level of information collected on them. They have no idea even how and to what extent this information is being shared with others. He is neither asked for approval. Most of the times data on personal information is collected from distant locations and it violate the mutual understanding between a person and the society. The decision makers appear to be outsider who take important decisions about the lives of individuals, the individuals affected remains unaware.

There are examples of violation of privacy and the citizens concern to stop the action. Lotus Corporation in 1991 started to sell a marketing aid by which detailed information on the personal and shopping habits of 80 million households (*i.e.*120 million Americans) was accessible to virtually anyone with a PC (Reitman 1991). This was aimed to boost up the small businesses, but might also invite nuisance like burglary, fraud *etc.* The company incurred a huge loss when it dropped this plan because of the likely consequences. The company also got a very bad name publicly. In another example, the Governments of Netherlands and West Germany in 1989 failed to alleviate or accommodate their citizens' concerns over privacy and the potential misuse of personal information and had to cancel national censuses.

In another interesting case, Dow Chemical in USA held the USA government responsible for violating privacy and their Fourth Amendment rights by showing close resolution aerial photography of their facility site. But it is also argued that limits on imaging technology or the scale of photography would be highly impractical and contrary to the development of science. Thus the context of privacy issues while using GIS should be dealt on a case to case basis rather than putting some guidelines or legislation that limits the use.

10.8 Environmental Modeling with GIS

There are there important interacting components required for successful natural resource and environmental management – policy, participation and information (Skidmore, 2002). Sustainable development of the natural resources is based on two critical factors, namely national, regional and local policy (legislations, sanctions, incentives, motivations *etc.*) and the participation by local people (farmers, land owners, stakeholders). The second factor requires timely, accurate and detailed information of land resources and their changes. This spatial information is provided and applied through GIS and remote sensing. Better spatial information means improved planning and decision making and generates harmony between production and conservation.

Ecology and environment seem to be the most important subjects/area where models along with GIS are being used most exhaustively. A model is an abstraction or simplification of reality (Odum, 1975; Jeffers, 1978; Duerr, 1979). It is because of its analytical capability that GIS is being increasingly used for decision making, planning and environmental management *e.g.* monitoring of deforestation, agro ecological zonation, ozone layer depletion, early warning system, monitoring of large ocean-atmospheric anomalies, wetland degradation, vegetation, land cover and soil resource mapping for input to the global climatic models. Models may be of many kinds- inductive/ deductive (based on logic), deterministic/ stochastic (based on processing method) and so on. Logical models, both inductive (models that derive a conclusion from a series of facts - exact mechanism not known) and deductive (draws a specific conclusion from a set of general prepositions – based on plausible physical laws, mechanism involved is also described) are used for environmental modeling; but inductive models mostly dominate spatial data handling. Global climatic model are a mixture of both deductive and inductive methods, whereas empirical/ numerical deterministic models applied to environmental sciences are calculated from (training) data collected in the field. Training data are those data that contain input and target values used for training or creating a model. In neural network study a network may be trained to look for certain patterns by providing it with a data set of known matching patterns. Training data is also used to assess the internal accuracy of a model. Inductive models also use training data sets – these are sometimes called inductive-empirical models *e.g.* USLE (a regression model) or BIOCLIM, CART (threshold empirical models – threshold values are also included in the models). USLE had been revised (using deduction methods) to include some more understanding of the (local) soil erosion process leading to RUSLE (Revised USLE), which a classic example of localized empirical model to be reapplied to new locations.

There are knowledge-driven models like *expert systems*, which has major four components – knowledge base of rules, method for processing the rules (*i.e.* the inference engine), and interface to the user and the independent *spatial* data stored in a GIS. Fuzzy systems and Bayesian expert system are some of the kinds of inductive knowledge-based approach.

The process-driven/ conceptual models (wherein mathematics is used to describe the factors controlling a process- can be *static* or *dynamic*) are mostly based on the deductive approach. These models are known to be a 'pure science product' as these describe a process based in understanding and established concepts (prepositions), through parameter values (Estimated from data). Most of the processes based models, as on today, are lumped model, but there has been substantial improvement in developing distributed models using raster GIS. Examples are TOPMODEL, SHE, ANSWERS (all are grid based). Geographical information systems help in linking the spatial and temporal elements of these models.

10.9 GIS and Watershed Management

The scientific and rational management of land and water resources for providing food and water to the millions of human and livestock populations is the need of the hour. Biotic and abiotic pressures over the years have resulted in land degradation, drastic reduction in biomass (the protective shield of land) with the loss of productive surface soils through surface runoff reducing the carrying capacity of streams and reservoirs. These along with widespread non-point source pollution from agricultural activities have taken the problem to a threatening dimension. On the other land, over-centralization of land and water management has failed over the years leading to the acceptance of watershed (a natural unit of eco-system planning and development)-based approach as the most rational approach today. By definition, 'watershed' is an appropriate geo-hydrological unit, in which rainfall occurring on the highest point of the area (ridgeline) drains at a common point. Watershed management programs include a broad domain of activities starting from the study of individual resources of a watershed for their potentials and constraints to finally study the possible exploration of these resources within the limits of tolerance, ensuing physical sustainability and economic viability. But to remember that the characteristics defining a watershed like climate (temperature, rainfall, evaporation demand etc), physiography, relief, morphometry, as well as socio-economic factors - are all interdependent and integrated in such a way that use of one resource may definitely influence others. This calls for a multi-phasic and integrated approach in watershed planning and management, which includes watershed characterization and monitoring to begin with.

GIS applications are beneficial in terms of many watershed management issues, locating possible sites suitable for groundwater recharge or rainwater harvesting may be just two of them. GIS with its ability to handle, retrieve, analyze and interpret a huge volume of data (virtually impossible to handle otherwise) offers a unique advantage over conventional methods. This provides a real-time integration of spatial information, fast and reliably.

The measurement and analyses of parameters defining a watershed require a wide range of information. Remote Sensing techniques providing a wide range of synoptic views and repetitive coverage over large areas have been proven to be an indispensable tool in this aspect. Remotely sensed data supplies a rich source of information about temporal and spatial changes of the earth surface conditions. The launching of the indigenous remote sensing satellites (IRS series) from 1988 onwards has made our country self-sufficient in utilizing this technology for soil and land cover mapping, monitoring of natural resources and for other purposes. The repetitive coverage provides an excellent opportunity to monitor land and water resources and to evaluate land cover changes through a comparison of images acquired for the same area at different times. The changes may be increased/decreased area under cultivation, change in surface water areas and water levels, mapping of drainage patterns with monitoring the change in river and drainage courses, reclamation of degraded land, erosion intensity mapping, monitoring changes in land covered with natural vegetation, fallow land, forest areas etc. The high-resolution satellite data presently available make it possible in mapping at 1:12,500 scale with detailed information. Assessment of the dynamics at the farm level is now possible with improving temporal resolution.

All these data effectively and successfully generated through satellite remote sensing is being used in GIS for analyses and interpretations. These are most needed for watershed monitoring and generating action plans for alternate land use systems. A long-term impact of the soil and water conservation programs implemented over a watershed thus becomes very accurate.

Apart from remotely sensed data, there are other ways of getting geographic data into a GIS. One can digitize physically existing documents, such as existing maps and drawings (like a soil or a drainage map), or can generate it electronically using a Global Positioning System (GPS). One can even get raw data from other sources and use in the system. Once the data is in the correct format, it's convenient to use and interpret according to one's own need.

The applications of GIS in watershed management can be broadly grouped into the following heads: (1) characterization of watersheds – the watershed characteristics like physiography and slopes, soils, geomorphology, morphometry, stream network, vegetation/biomass, land use/ land cover, surface water etc. can be quantified and assessed; (2) monitoring – spatial and temporal changes can be effectively monitored and prioritization becomes easy; (3) Estimation and/or prediction of runoff and soil loss in watershed – GIS can be used with various runoff and soil erosion equations for their quantification; (4) Identification of potential water-harvesting sites – this identifies where and how much water can be harvested, which is absolutely essential for designing water harvesting structures; (5) Identification of potential soil erosion prone areas/ zones in a watershed - for designing soil conservation structures (agronomic or engineering –as required); (6) Land suitability classification – identification of different parcels of land for sustainable utilization based on their potentiality; (7) Water balance studies and ground water prospecting; (8) Mapping of wasteland, saline/alkali soils and other degraded land; (9) Assessment of overall health of a watershed and finally (10) Making, evaluating and implementing any action plan for watershed management. A GIS database provides decision-makers with a comprehensive visual and tabular means for analyses on which to construct and support decisions.

A Govt. of India programme named Integrated Mission for Sustainable Development (IMSD) has been operational in 175 districts in our country (all together 84 million hectares of area). The basic objective of this program, which uses GIS environment for integrating and analyzing land, water, vegetation, meteorological and socio-economic data, is to generate locale-specific recommendations for sustainable development of land and water on watershed basis. It also aims in implementing the recommended action plans on the ground.

GIS could shape the future of the field of watershed management in that it allows the managers to provide communities with the tools to be informed of their water (shed) situation, and to realize the impacts of various actions. Citizens will be able to make decisions and take actions toward maintaining and monitoring a productive watershed system. With the rapid development of software in the field of GIS, we are entering an era in which the use of GIS can better organize and integrate scientific data, address spatial and temporal variability, model soil and landscape using quantitative and statistical methods and interpret the problems in more quick and reliable way.

STUDY QUESTIONS

1. Define a Geographic Information System. What are its central concepts? How far a GIS is superior over other conventional systems?
2. What are the principal components of GIS and which one is the most important among them?
3. How has the remote sensing system started a new era in the field of GIS? Enumerate the Indian experience and capabilities in remote sensing.
4. What are the major areas of GIS applications?
5. What is cartography and why is it considered as the root of GIS?
6. Explain the term 'geoid'? How does it differ from other models of the shape of the Earth?
7. What are coordinates? How many coordinate systems do you know and which are used to locate a place on the Earth?
8. What is UTM and why is it so popular? How MGRS differs from UTM?
9. What is a datum? How local datum is different from universal/global datum?
10. What is map projection? Explain some ways of classifying the Earth projection systems? What are the steps in map projection?
11. Explain perspective map projection with examples.
12. Enumerate cylindrical, conic and azimuthal projection systems.
13. What is the criteria in selecting a suitable projection system for one's GIS project?
14. What are the steps in developing a digital representation of the real world phenomenon in GIS?
15. What are vector and raster data model? How they are pictorially represented? List some prominent raster and vector data formats?
16. Which one between vector and raster would you prefer for your analysis in GIS?
17. What is topology and how is it built up?
18. Explain the terms 'rasterization' and 'vectorization'? Why they are used?
19. How does an analog map differ from a digital map?

20. Explain in detail various data sources for GIS, their availability and potential use.
21. What are the devices commonly used for GIS data input?
22. What is digitization? List the possible errors in digitization.
23. What is GPS? Explain the possible applications of GPS in GIS? Enumerate its various components and the precautions which should be taken so as to minimize errors in using it.
24. What is a database system? Explain the hierarchical, network, relational and object-oriented database models with their comparative advantages and disadvantages.
25. How data are efficiently and effectively organized in a raster system?
26. What is a quadtree structure?
27. What is spatial analysis? What are the major spatial functions available in GIS?
28. How least cost path is calculated in GIS?
29. How a query is made in GIS? What are the attribute based operations in GIS?
30. What is buffering analysis in GIS? How does it differ from proximity analysis?
31. Explain with examples various types of network analyses in GIS.
32. Show (with figures) the raster and vector overlay functions in GIS.
33. What are the various methods of interpolation? What is the most common interpolation technique performed in GIS?
34. What is artificial neural network and how is it useful in spatial analysis using GIS?
35. What are the various kinds of output in GIS?
36. Explain the principles of remote sensing with special emphasis on the use of different portions of EM spectrum of sunlight.
37. How are spatial, spectral, radiometric and temporal resolution defined in remote sensing system?
38. How the remote sensing data is possibly integrated in GIS? Explain with examples.
39. What are various software options in remote sensing?
40. Explain the role of GIS in the Internet era.
41. How the geographical objects with indeterminate or fuzzy boundaries are addressed by GIS?
42. What are the possible legal and privacy issues related to the use of GIS?
43. What could be the possible errors in GIS?
44. Explain the possibility of GIS or Geoinformation science to be inducted in school education?
45. Describe the potential role of GIS in ecological modeling and watershed management.

Glossary

Absolute accuracy A measure of the difference between the location of an object as specified in a GIS, and its true location in the real world.

Access path A description of the hierarchical chain of directories that define the logical location of a file.

Access time A measure of the time interval between the instant that the data are called from storage and the instant that delivery is complete.

Accuracy Conformity with a standard or correctness in measurement. Accuracy relates to the quality of a result and is distinguished from precision.

Address (1) A number referring to a location in computer memory, 2) A name that identifies a location on a computer network so that other machines can communicate with it.

Address geocoding Process of assigning alphanumeric locational identifiers (such as the municipal address or physical location) to spatially related information. For example, an address may be matched to an address range on a street segment, or a given spatial area (i.e., the limits of a polygon, a line segment, a point along the segment, or an absolute point that has been coordinated).

ADMATCH (address matching) A computer program that can match street addresses with census tracts and/or block codes and convert street addresses to coordinates for computer mapping.

Aerial photograph A photograph taken vertically downward from the air. Aerial photograph images may be in the form of paper prints, often 9" x 9", or transparent film.

AGIS A GIS marketed by Delta Data Systems, Inc. It has both raster and vector data acquisition and attribution capabilities in a single operating environment. The user can link dBASE III to its relational database.

AIX A UNIX system variant developed by IBM for use on IBM minicomputers and workstations.

Albedo The ratio of the light reflected by any object (like surface of Earth or atmosphere, a water body, a leaf, a soil element etc.) to that received by it.

Algorithm Statement of the steps to be followed in the solution of a problem.

Aliasing The appearance of jagged lines on a raster display.

Alphanumeric Consisting of both letters and numbers, and possibly including other symbols such as punctuation marks.

AM/FM (**A**utomated **M**apping/**F**acilities **M**anagement) The use of a computer-aided mapping or geographic information system for public works and utility information management and similar applications. Software specifically developed for AM/FM generally lacks spatial analysis and COGO functions.

Analog Information stored and processed with signal intensity or other measurement of a continuous physical variable (field and laboratory measurements of temperature, pressure, voltage, radiation, etc.). Analog information can be converted to its digital equivalent.

Analog/Digital conversion The process of converting data from analog to machine usable digital format via processes such as digitizing or scanning (See also Digitize Scanning).

Anti-aliasing The process of removing or smoothing the jagged, stair-step appearance of a digital line by filling in some of the intermediate and flanking cells in lower-intensity colors (also see Aliasing).

Arc A line established by connecting a set of points. It has length but no area and often serves as one side of a polygon. Arcs may begin and end with nodes and have points of inflection defined by vertices. An arc usually represents a continuous common boundary between two adjoining mapping units.

Arc/Info A vector-based GIS developed and marketed by ESRI, Inc.

Area (1) A polygon that encloses a homogeneous unit (e.g., lake, land unit, district, state, country), (2) A level of spatial measurement that references the size or extent of a two-dimensional defined space.

Arithmetic operator In a GIS, operations performed on attributes of objects or entire maps, such as add, subtract, multiply, and divide.

Array A data structure that treats continuous element in the form of a grid or matrix as a series of addressable repeating patterns. Each element in an array is referenced by an index which gives the location of the element in relation to other elements.

ASCII (American **S**tandard **C**ode for **I**nformation **I**nterchange, pronounced "askee") The 256 characters that comprise a computer's alphabet. The character set encoded into digital values between 0 and 127 includes the lowercase and uppercase letters, the numerals 0-9, punctuation marks, special symbols (such as @#$%^&*), and non-displaying characters often used as printer control codes. Values from 128 to 255 are not standardized, and are not part of the basic ASCII code. However, most PCs use a common extended ASCII character set for values from 128 to 255. (Also see ASCII file).

ASCII file A text-only file. Documents in most word processors are not text-only files, since they include coded header information and formatting characters. However, most word processors import ASCII files and have an Export, Save/Text-only, or print-to-file utility that converts a document into an ASCII format. (Also see ASCII).

Aspect Horizontal direction in which a slope faces. Commonly expressed in degrees clockwise from north (e.g., a southeast facing slope has an aspect of 135 degrees). (See also slope)

Aspect ratio The ratio of horizontal scale to vertical scale for printing or display. For graphics and image processing, square cells/pixels are best (aspect 1:1). Some display devices have a non-square aspect, which causes images to appear stretched or distorted. Standard broadcast video has an aspect ratio of 4:3, which must be corrected in any frame-grabbing or other digitization process.

Assembly language A low-level (primitive) programming language that uses mnemonics rather than English-like statements.

Association A variety of relationships among two or more entities.

Attribute A numeric and/or text description of a spatial entity (e.g., attributes of a land parcel might include address, owner's name, and other property values). Attribute data is typically stored in tabular format.

AutoCAD A commercial Computer Aided Design (CAD) software package, developed and distributed by Autodesk, Inc.

Autocorrelation Statistical concepts expressing the degree to which one value of an attribute (co)varies with other values of the same attribute. For GIS, this refers to the degree to which the values of an attribute of two objects (co)vary with a separating distance between them in space.

Automated cartography Drawing maps with aid of computer-driven devices like plotters and visual display units (screens).

Automated Mapping/Facilities Management (AM/FM) See AM/FM.

AVHRR imagery **A**dvanced **V**ery **H**igh **R**esolution **R**adiometer imagery produced by NOAA satellites.

Azimuth: The horizontal direction of a vector, measured clockwise from north.

Azimuthal projections A class of map projections where directions of all lines radiating from a central point are the same as directions of the corresponding lines on the sphere. This projection is formed onto a plane, which is usually tangent to the globe at a pole (a polar projection), at a point on the equator, or at any selected intermediate point.

Band A range of wavelengths of electromagnetic radiation (Also see spectral band).

Base layer A layer containing a variety of features often used for locational reference (e.g., section corners, political boundaries, and other major features) or to establish geodetic control that ensures geographic fidelity between data sets in the GIS.

Base map Any map containing geographical features used for locational (map grid) reference. These provide the background on which other data layers can be overlayed and analysed.

Batch processing The collection of programs in a queue for further processing, with priorities assigned according to estimates of the processing time required.

Benchmark A reference point for measurements, normally one that is used for elevation reference.

Benchmark test The procedure to evaluate the capabilities of a computer system (software+hardware) in terms of performance and ability to meet customer requirements.

Best Linear Unbiased Estimate The product from an interpolation function, which was optimized with chosen interpolation weights, of the value of a variable at a given point.

Bicubic spline See B-Spline.

Bilinear interpolation A mathematical method for interpolating a new cell's value within a 2×2 neighborhood of cells (Also see Interpolation).

Binary A number system with base 2 that uses only the data values 0 and 1. Each place represents a power of 2 e.g., the decimal number 13 is represented as 1101 ($1 \times 8 + 1 \times 4 + 0 \times 2 + 1 \times 1$).

Binary file A file containing characters that are in machine-readable form.

Bit The smallest unit of information that a computer can store and process digitally. A bit has two possible values, 1 or 0, which correspond to 'yes' and 'no' (or 'on' and 'off') (also see Byte).

Bitmap An image stored as a pattern of dots (or pels).

Boolean operator Search strategy for information retrieval based on the use of the logical operators AND (union), OR (intersection), NOT (complement), and XOR (exclusion).

BPI (Bits Per Inch) Density of bits recorded on a magnetic tape: 800, 1600 and 6500 are the common standards.

B-Spline The sums of other splines that by definition have the value of zero outside the interval of interest, therefore, allow local fitting from low-order polynomials in a simple way. B-Splines are often used for smoothing digitized lines to display, such as the boundaries on soil and geological maps.

Buffer An area of specified distance (radius) around a map feature (point, line or polygon) (Also see proximity analysis).

Byte The data storage unit composed of 8 bits having 256 values (0 to 255). A collection of bytes (often 4 or 8 bytes) represents real numbers and integers larger than 255. In text-oriented processes, each byte represents one character of text. In 8-bit raster processes, each byte represents one cell value and may correspond to one grayscale pixel on the image display.

CAD See Computer Aided Design.

Cadastre A survey involving mapping, tracing and recording of public lands and private estates for tax and legal purposes.

Calibrate (geographic) To bring a raster or vector object into alignment with some geographic coordinate system.

CAM – See Computer Aided Mapping

Capture To freeze and digitize a standard video input signal (such as VHS tape or broadcast video).

Cardinal direction The four principal directions: North, South, East and West.

Cartesian coordinates A plane coordinate system in which the locations of points in space are expressed by reference to two or three planes, called the 'coordinate planes' (x,y or x,y,z).

Cartography The art or science of making maps.

Cascade A way of arranging open windows on the screen so that they overlap each other, with the title bar of each window remaining visible.

Cathode Ray Tube (CRT) An electronic screen for displaying information or graphics.

Cell size The dimensions of the area on the ground to which a raster cell value applies (a cell size of 30 meters signifies that the value in each cell of the raster object applies to a 30 × 30 meter area in the study area).

Cell The basic element of spatial information in the raster (grid) description of spatial features. (Also see pixel).

Centroid The centre of a polygon. In the case of an irregularly shaped polygon, it is roughly the equivalent of its centre of gravity; used in GIS to index the polygon within which they are located.

Chain coding A method of reducing the storage requirements of a raster.

Channel (GPS) A part of GPS receiver consisting of the circuitry necessary to tune the signal from a GPS satellite.

Character An alphanumerical (alphabetical or numerical) value that represents a single unit of data.

Choropleth map A map in which areas of different value are separated by clearly defined boundaries. The value of the underlying data (e.g., soil moisture) are represented by colour or shading densities, and the map legend acts as a look-up key to explain the values shown on the map.

Classification Grouping cells (often by spectral characteristics) from one, or more characteristically, from a set of co-registered raster objects to isolate and name image features (crop varieties, wetlands, forest, or other surface cover).

Clip The process of extracting a portion of a coverage or map from a larger coverage, much like a cookie cutter.

Cluster analysis Reduction of the complexity of the data set by assigning objects to classes in a classification system.

Color composite In satellite imagery, a color negative, transparency or print produced by allowing the reflectance recorded for each band of a multi-spectral image to be represented by a proportionate intensity of one of the primary colors.

Column The vertical dimension of a table, representing the variables of the geographical features listed in the rows.

Composite map A single map created by joining together several separately digitized or scanned maps.

Computer aided design (CAD) Software with the capability of performing standard engineering drawing and architecture design functions. Unlike GIS, the database is generally not structured to allow spatial analysis.

Computer aided mapping (CAM) Software with the capability of performing standard mapping functions, typically using a vector format. Unlike GIS, it cannot analyze or process the database.

Conditional operator An SQL (Structured Query Language) keyword specifying how to compare values (e.g. = , >, Like and Containing).

Connectivity A topological property relating to how geographical features relate to one another spatially.

Connectivity analysis See network analysis.

Contiguity analysis See neighborhood analysis.

Continuous data Raster data represented by a three-dimensional surface so that such intermediate values can always be derived, e,g. an elevation raster has continuous data because an elevation of 400 and an elevation of 500 can fairly be averaged to assign an intermediate elevation of 450.

Contour An imaginary line on the ground, all points of which are at the same elevation, usually expressed as a length or elevation above some datum such as mean sea level.

Contour lines Lines joining points of equal elevation.

Contour map A topographic map that uses contour lines to portray relief.

Control point A permanently fixed point on the ground, the location of which has been accurately determined and recorded for reference and future use.

Coordinate Used to represent location on the earth's surface relative to other locations in either two (x,y) or three (x,y,z) dimensions.

Coordinate geometry (COGO) A data input method to convert survey data into geographic locations.

Cordinate system A kind of reference system, which use linear or angular quantities to designate the relative position of points within that particular system.

Coverage The graphic and attribute data related to a particular data-theme for a study area. In GIS a coverage represents a single layer or theme.

Dangle A line with one end (or both) that does not connect to any other feature in the map. This results from faulty digitization.

Data Facts about real-world entities, organized for analysis. Data of direct relevance to GIS include attribute information that describes the characteristics of the entity; geographical information that describes the position of the entity in space or in absolute relative to other things in space.

Data capture Operations required to encode data in a computer readable digital form (e.g., digitizing, scanning). See also data input.

Data conversion Translation of data from one storage format or media to another for the purpose of transferring it from one GIS to another; also refers to the process of transforming maps from manual to digital form.

Data Definition Language (DDL) Repository of information about the definition, structure, and usage of data; does not contain the actual data.

Data element A specific item of information appearing in a set of data.

Data input (data capture) Entering data into a computer; geographic data is generally entered into a GIS database via a digitizer or a scanner.

Data layer Refers to data having similar characteristics being contained in the same plane or overlay (e.g., roads, rivers).

Data model A user-defined, abstract representation of data describing the behavior of the geographical entities represented by the data.

Data set A collection of logically related information arranged in a specified manner.

Database A logical collection of files managed as a unit. A GIS database includes data about both the position and the attributes of geographic features.

Database management system (DBMS) A systematic approach to maintaining, accessing, reporting, and analyzing attribute data. GIS packages may use DBMS to handle some data management tasks.

Database structure The physical organization of data elements assigned to files and relationships among files.

Datum A base or reference elevation used as the origin to define subsequent elevation. Often refers to mean sea level; also singular of data.

Degree Unit of measurement equal to 1/360th of a circle.

Digital data Collection of similar and related data records that are converted into a form for use by a computer.

Digital elevation model (DEM) A data file of a topographic surface arranged as a set of x,y,z coordinates where z represents surface elevation. It is the digital equivalent of the elevation data portrayed on a topographic base map.

Digital exchange format (DXF) A standard format developed by Autodesk Inc., for exchanging digital cartographic (map) files and associated spatial data between different GIS.

Digital line graph (DLG) A file or data structure from the US Geological Survey (USGS) that includes digital non-topographic information from the USGS map base categories such as transportation, hydrology, and public land survey boundaries.

Digital terrain model (DTM) See digital elevation model.

Digitize A process used to create digital data from non-digital or analog data (Also see data conversion).

Digitizer A device consisting of a table and a cursor (often with cross-hairs and keys) that is used for capturing and recording the locations of map features as x,y Cartesian coordinates.

Directory A set of data files, stored on computer disk The location of a directory is specified with a path name.

Discrete data Geographical features (point, line or polygon) well-defined by boundaries.

Display Graphic presentation of the output data of any device or system.

Dissolve The process of removing boundaries between adjacent polygons having the same values for a specified attribute.

Distance analysis Geographic computations based on Euclidean and/or non-Euclidean distances between features.

Dot matrix plotter A plotter, of which the printing head consists of many, closely spaced (100-400 per inch) wire points that can write dots on the paper to make a map. Also known as an electrostatic plotter or matrix plotter.

Dots Per Inch (DPI) Resolution of graphic displays and printers, representing the number of pixels per inch.

Drum plotter A device with a rotating cylindrical drawing surface and paper reels for plotting graphic images on a continuous roll of paper.

Easting One of two geographic coordinate values (the other being northing) used to specify the exact location of a feature on a map.

Editing The detection and correction of errors in GIS.

Electromagnetic spectrum The entire continuum of spectrum of all kinds of electromagnetic and visible radiation, from gamma rays having a wavelength of 0.001 angstrom to long waves having a wavelength of more than 1 million km. (Also see Spectral band).

Element (Map element) Basic geographical unit of information, such as a point/node, line/arc, area/ polygon or pixel/cell. Also known as map entity.

Ellipsoid An ellipse rotated about its shorter axis; Earth is an irregular ellipsoid. Recently, satellite data has been used to construct ellipsoid models that relate coordinate measurements to the Earth's center of mass.

Encode Convert data to a form that is suitable for entry into a computer.

Entity (Map entity) Object or feature about which information is stored. The information describes where the entity is (location - stored as points, lines, or polygons) and what the entity is (identity - e.g., lake, tree, house). (Also see Element).

ERDAS (Earth Resources Data Analysis System) A raster-oriented microcomputer image processing and GIS system marketed by ERDAS, Inc.

Euclidean distance The shortest distance joining two points in the plane.

Expert system A computer system that can perform at or near the level of a human expert.

Extrapolation Process of extending results of sampling beyond the extent of sample points.

False easting An adjustment constant added to easting coordinate values to eliminate negative numbers. (Also see Easting).

False northing An adjustment constant added to northing coordinate values to eliminate negative numbers. (Also see Northing).

Feature A geographic component of the earth's surface that has both spatial and attribute data associated with it (e.g., well, road, lake).

Field A common synonym for column in a database.

File A collection of related records (textual and graphical) in a computer treated as a unit.

File Transfer Protocol (FTP) A common protocol for copying or moving data across a computer network or the Internet.

Filter Mathematically defined operation, basically a grid of weighted pixel values, used to remove long range (high-pass) or short range (low-pass) variation, or to generalize spatial patterns from a raster data set.

Flow accumulation In a raster-based watershed analysis, the number of cells that drain into some selected cells.

Flow direction In a raster-based watershed analysis, each cell is assumed to drain into one of its eight neighbors (left, right, up, down, plus the four diagonals).

Fourier transform (Fourier analysis) Two-way frequency transform for identifying and removing unwanted spatial frequency components in an image; normally used to identify and remove systematic noise spikes, such as regular lines that may have been introduced by a faulty image collection device.

Geocoding The process of creating an x,y coordinate location from another geographic location description, such as an address. The term evolved from geographic coding.

Geographic coordinates Values specifying the location of features in a standard, absolute worldwide coordinate system (e.g., latitude/longitude, state plane coordinates, Universal Transverse Mercator).

Geographic data Data that convey the locations and descriptions of geographic features (see also spatial data).

Geographic Information System (GIS) A computer system designed to collect, manage, analyze and display large volumes of spatially referenced and associated attribute data.

Geographically referenced data See Geographic data.

Geometric correction Alters data to correspond with true ground or image space in a known coordinate system.

Georeferencing A process of referencing points on the surface of the Earth to points on a map. Examples of referencing systems include latitude/longitude, universal transverse mercator, and state plane coordinates.

Geodesy (also geodetic) Science of the size and shape of the Earth.

Gigabyte (GB) A computer unit of measurement for 10^9 bytes.

Global positioning system (GPS) A system that uses satellites to locate positions on the Earth's surface.

Gnomonic projection A type of perspective azimuthal map projection where the point of the projection is the center of the sphere.

Graphical User Interface (GUI) A user interface that uses such objects as icons, menus, pointers, and scroll bars to allow the user to choose various options and run programs using graphical interactions such as moving, pointing, and clicking.

GRASS (**G**eographic **R**esources **A**nalysis **S**upport **S**ystem) A public domain raster-based GIS system designed for minicomputers.

Graticule Network of parallels and meridians on a map or chart.

Grid A data structure that uses rectangular units or grid cells arranged in rows and columns to represent map features; also known as raster.

GRID A raster module in the Arc/Info GIS package.

Ground control point See control point.

Ground truth To verify the correctness of remote sensing information by use of ancillary information such as field studies.

Hard copy An analog image of a map or diagram (e.g., a paper map produced by a printer or plotter) generated from a digital data set.

Hierarchical database structure A method of structuring data or other information so that the units of data storage are connected by a hierarchically defined pathway.

High-level language A computer programming language using command statements, symbols, and words that resemble English-language statements e.g. FORTRAN, PASCAL, C, COBOL, BASIC.

Hypsometry The measurement of the elevation of the Earth's surface with respect to sea level.

I/O address Locations within the input/output address space of computer used by a device such as a printer or scanner; used for communication between software and a device.

IDRISI A public domain, raster-based GIS system from Clark University.

Image processing (also image analysis) Computerized routines for information extraction (e.g., pattern recognition, classification) from remotely sensed images to obtain categories of information about specific features.

Info The database program used by Arc/Info. (Also see Arc/Info)

Ink jet plotter/ printer – An output device that creates images by projecting a jet of ink onto the paper.

Input Data that have been entered into a computer system; to enter data into a computer.

Input device A hardware component for data entry.

Interpolation The process of estimating the value of an unsampled data point for a given x,y,z location based on the values of surrounding sampled data points.

Island polygon A separate, individual polygon such as a lake.

Isopleth map (Also Isoline map) A map displaying the distribution of an attribute in terms of lines connecting points of equal value (e.g., a contour map).

Join To connect two or more separately digitized maps.

Kilobyte (or KB or K) A unit of measure equal to 1024 bytes (2^{10}), which is taken as 1000 bytes approximately (Also see megabyte, gigabyte and terabyte).

Kriging An interpolation technique for obtaining statistically unbiased estimates of surface elevations from a set of control points.

Label A unique name or code assigned to identify the geographic representation of a map feature.

Labeling The process of assigning unique labels to polygons.

Land Information System (LIS) A special type of GIS that manages and analyzes data related to land ownership (e.g., tax parcels, urban infrastructure, property assessment). A GIS used for municipal or country level applications is typically structured as an LIS. Data stored in an LIS are commonly recorded very accurately on a large-scale map (e.g., at scales of 1:1,000 to 1:10,000).

LANDSAT The generic name for a series of natural resource scanning satellites launched by the United States since 1972.

Large-scale A map scale that covers a relatively small area on the ground and has a high level of details.

Laser printer/ plotter An output device in which the information is written onto light-sensitive material using a laser.

Latitude A system of referencing relative north-south locations on the earth's surface, measured in degrees, minutes, and seconds north or south of the equator to the poles.

Layer A logical separation of mapped information representing a theme (e.g., roads, soils, vegetative cover).

Legend The part of the drawn map explaining the meaning of the symbols used to code the depicted geographical elements.

Line In GIS, a set of ordered coordinates that represent the shape of a linear geographic feature; has length and direction but no area; e.g., streams, roads, telecommunication lines etc.

Line on polygon overlay To superimpose a set of lines on a set of polygons, breaking the lines at intersections with polygon boundaries.

LIS See Land Information System.

Longitude A system of referencing relative east-west locations on the earth's surface, measured in degrees, minutes, and seconds east or west of the 'Prime Meridian' which runs through Greenwich, England.

Look-up table An array of data values that can be quickly accessed by a computer program to convert data from one form to another.

Machine language Instructions coded so that the computer can recognize and execute them.

Map A two-dimensional graphic representation of the earth's surface that displays spatial relationships among features.

Map algebra Process of applying mathematical operations and specify mathematical relationships to map themes to generate new map.

Map analysis See Spatial analysis.

Map extent The rectangular limits (x-min, y-min, x-max, y-max) that include all the features displayed on a graphics display terminal or on a plotted map.

Map generalization The process of reducing details on a map; often in conjunction with reduction in map scale.

Map projection The geometrical or mathematical technique to represent all or part of a rounded surface (Earth) on a flat surface.

Map scale The relationship that exists between a distance on a map and the corresponding distance on the Earth.

Map theme See theme.

MapInfo A vector-based desktop GIS marketed by MapInfo Corporation; operates on DOS, Windows, and Unix, and can be linked to dBASE databases.

Mapping unit A set of areas drawn on a map to represent a well-defined feature or set of features.

Megabyte (Mb) A unit of measurement for 10^6 bytes.

Mercator projection A map projection centered along the equator with evenly spaced meridians perpendicular to parallels spaced progressively farther apart poleward so that compass bearings are not distorted.

Merge To combine two or more maps or data sets together into a single, coherent map or database without redundant information.

Meridian A great circle on the Earth's surface that passes through the terrestrial poles.

Metadata The information about the nature of data items (accuracy, extent, copyright and restrictions etc.) is called metadata; not actual data, but 'data about data'.

Minute The sixtieth part of a degree of angular measurement, often represented by the sign '¢', as in 30¢, read '30 minutes'.

Model A set of rules and procedures that represent a view of reality for conducting spatial analysis to generate a result.

Module A separate and distinct piece of hardware or software that can be connected with other modules to form a system.

MOSS (Map Overlay and Statistical System) A pioneer vector-based geographic information system (GIS) developed and still widely used.

MSS (Multi-Spectral Scanner) A sensing device on the LANDSAT satellite that collects simultaneous images over multiple ranges of the spectrum.

Neighborhood analysis Analytical technique to determine whether a set of areas (polygons) are situated next to each other and to examine their interrelationships.

Network analysis Technique used to conduct analyses on a set of points (nodes) and lines that are connected to each other.

Network database structure A method of arranging data in a database so that explicit connections and relations are defined by links or pointers of a many-to-many type.

Node A point at which two or more arcs or lines meet; also the distinct point in a feature where lines begin and end; must in vector GIS to define topology.

Noise Irregular variation, usually short range that cannot be easily explained or associated with major mapped features or processes.

Non-graphic information See Attribute.

Northing A rectangular (x,y) coordinate measurement of distance north from an east-west reference line, usually the Equator or other parallel used as the axis of origin for a map zone or projection.

Nugget variance The nugget variance is a spatially uncorrelated noise (the residual variances of measurement errors + spatial variations that occur over distances much shorter than the sample spacing.

Object A single, identifiable unit or entity.

Object-oriented database The intrinsic representation and storage of real world entities in a GIS database rather than the use of cells, points, lines, or polygons to store the object.

Optimal path selection Process of using a network database of roads and road intersections to select the optimal path based on specified time/distance relationships.

Output The final product of a GIS in the form of a printed map, screen displays, a tabular data summary, or a data file.

Overlay analysis The process of combining spatial information from two or more maps of the same geographic area to derive a map consisting of new spatial boundaries and entities or themes.

Pan To widen the view perspective in a graphic display (i.e., "zoom out").

Parallels Circles on the Earth's surface, or lines on a map that are perpendicular to the axis of the Earth, and mark latitude north or south of the equator.

Pen plotter A plotter that uses pens as drawing elements.

Photogrammetry Science of obtaining precise measurements from images.

Pixel (Picture **e**lement) The smallest discrete element which makes up a raster image; also refers to individual grid cells in a raster.

Plane coordinates Coordinates specifying the locations of points in a plane.

Planimetric map A map that represents only the horizontal positions of features, typically used as part of a base map.

Point A single x,y coordinate that represents a geographic feature that is too small to be displayed as a line or a polygon at a particular scale (wells, weather stations, sampling sites etc.).

Polygon A two-dimensional figure with three or more sides intersecting at a like number of points. It is defined or bounded by a closed line or arc and has attributes that describe its geographic features.

Precision The degree of discrimination with which a quantity is stated. For example, a three-digit number discriminates among 1,000 possibilities.

Preprocessing Preliminary processing or transformation of raw data required to facilitate further cartographic processing.

Projection See Map projection.

Proximity analysis See Network analysis.

Quadrant A quarter of a circle, 90 degrees.

Quadtree A hierarchical data model that provides a compact raster representation by using a variable-sized grid cell. Finer subdivisions are used in areas require finer detail, providing a higher which level of resolution.

Query To find specific information in a database.

Query language A high-level programming language, using words that closely resemble ordinary language that provides for the easy retrieval of information.

Raster algebra Manipulations and functions that operate on raster objects cell by cell.

Raster data Cell data arranged in a regular grid pattern in which each unit (pixel or cell) in the grid is assigned an identifying value based on its characteristics.

Raster/vector conversion (Vectorization) To convert data from raster to vector format with position and orientation selected by the user.

Rasterization – See Vector/raster conversion.

RDBMS See Relational database management system.

Reclassification Procedure to change the classification of existing data.

Record A group of logically related attribute descriptors for one feature (e.g., a house in an attribute table that contains information about houses in a community). It is commonly displayed as a row in a relational attribute table.

Registration Process of geometrically alignment of different themes to a common coordinate system or to one theme that is used as a standard so that themes can be consistently overlaid.

Relational database management system (RDBMS) Any database management system that can access a relational database, generate responses to adhoc queries and preformatted reports by using relational algebra or calculus, and perform routine update operations.

Relational database structure Database structure in the form of records so that relations between different entities and attributes can be used for data access and transformation.

Relational join The operation of relating and physically merging two attribute tables using a common or relational item.

Relational operators Phrases such as greater than, less than, maximum, minimum, and contains that are used to compare values associated with spatial data.

Remote sensing The act of detection and/or identification of an object, series of objects, or landscape without having the sensor in direct contact with the object. Often results in the generation of image data.

Resampling A process of assigning values to new, rectified, or rescaled (re-sized) cells in a raster database.

Resolution The minimum distance between two objects that can be distinguished by a sensor- often a synonym for spatial resolution, but also applies to spectral and temporal aspects of remote sensing imaging systems.

Routing analysis A type of network analysis that is used to optimize vehicle routing (*e.g.*, routing of emergency services, school buses, mail delivery, municipal garbage collection).

Run-length coding A method of reducing the storage requirements of a raster.

Scale The ratio or fraction between the size of an object on a map and its size in the real world. A scale represented as 1:25,000 means that one unit of distance on the map represents 25,000 of the same units of distance on the earth.

Scanning Process to convert analog information from maps, photographs, or overlays into a digital format usable by a computer.

Semivariogram A graphical representation relating the variance of the difference in value of an attribute at pairs of the sample points to separation distance.

Sliver A gap or overlap that is generated by combining two or more coverages that are not coincident or perfectly conflated.

Slope The rate of rise or fall of a quantity against horizontal distance. It may be expressed as a ratio, decimal, fraction, percentage, or the tangent of the angle of inclination.

Small scale A map scale that covers a relatively large area and has generalized labels.

Smoothing The reduction of the local variability of data and, when applied to a spatially distributed variable, results in a reduction of local variance.

SPANS A GIS marketed by TYDAC Technologies Corporation. It has raster, vector, and quadtree data structures, and operates on PCs and workstations as a self-contained desktop GIS.

Spatial Refers to phenomena distributed in space and therefore having physical dimensions and geography.

Spatial analysis Analytical techniques associated with the study of the location of geographical entities together with their spatial dimensions.

Spatial data Data pertaining to the location, shape, and relationships among geographical features. These can be classified and stored as point, line, area, polygon, grid cell, or object.

Spatial database A collection of spatial information related by common facts or themes.

Spatial decision support system (SDSS) A customized computer-based information system that utilizes decision rules and models and incorporates spatial data.

Spatial resolution See resolution.

Spectral band (or spectral region) A well defined, continuous wavelength range in the spectrum of reflected or radiated electromagnetic energy (Also see Electromagnetic spectrum).

Spline An interpolating polynomial for a set of coordinate points used to fit a curve that connects the points (Also see B-spline).

SPOT The generic name for a series of natural resource scanning satellites launched by France.

Stereoplotter A machine which, when loaded with a pair of overlapping aerial photographs, can be used to determine precise location and altitudes of objects that appear in the photographs.

Tablet (or digitizer tablet) A small digitizer used for interactive work on a graphics workstation.

Tabular data Data (usually attribute) organized into logical tables. Tables contain items and records or rows and columns.

Terabyte (or Tb) A unit of measurement for 10^{12} bytes.

Terrain analysis Analytical techniques that quantify terrain parameters (slope, aspect) or the effect of terrain on a particular operation.

Tesselation The process of splitting an area into small, manageable units or subareas (tiles or rasters).

Thematic map A map related to a single topic, theme, or subject of discourse; also called topical, geographic, special purpose, distribution, parametric, or planimetric maps.

Theme The overall topic of a map layer in which the spatial variation of a single phenomenon is illustrated (e.g., a vegetation theme map might illustrate vegetative areas such as crop land, fallow, forest, scrubs etc).

Thiessen polygons Methods of dividing areas into tiles according to minimum distance between sample points. Also known as Dirichlet tesselation or Voronoi polygons.

Tics Geographic control points representing known or identifiable locations on the earth's surface (e.g., longitude, latitude). Also called ground control points.

TIGER (Topologically Integrated Geographic Encoding and Referencing file) The nationwide digital database of planimetric base map features developed by the U.S. Bureau of the Census for the 1990 Census.

Topography The shape of the surface of the Earth in a given area.

Topologically linked database A database containing linked forms of digitized polygons and non-spatial attributes.

Topology (or Topological relationship) The spatial relationships between connecting or adjacent coverage features (e.g., points, lines, and polygons).

Trend surface analysis The simplest way to describe gradual long-range variations is to model them by polynomial regression. The idea is to fit a polynomial line or surface, depending on whether the data are in one or two dimensions, by least squares through the data points. It is assumed that the spatial coordinates (x,y) and their powers and products are the independent variables, and that z, the property of interest, is the dependent variable.

Triangulated irregular network (TIN) A data structure that describes a three-dimensional surface as a series of irregularly shaped triangles.

Tuple See Record.

Universal transverse mercator (UTM) grid A system of plane coordinates based upon 60 north-south trending zones, each 6 degrees of longitude wide, that circle the globe. Used to derive geographic coordinates, normally in meters, east and north of an origin that is defined uniquely for each zone.

UNIX A modern, general purpose computer operating system.

Vector data Data comprised of x-y coordinate representations of locations on the earth that take the form of single points, strings of points (lines or arcs), or closed lines (polygons).

Vector/raster conversion (also Rasterization) To convert data from vector or point formats to raster format with grid cell size, position, and orientation selected by the user.

Vectorization See raster/vector conversion.

Vertex (pl. vertices) Intermediate points along a line curve, or arc. They represent the critical points of inflection along the arcs, thereby reflecting its shape.

Viewshed The boundaries of sight from a single vantage point, assuming an unobstructed surface (disregard trees and buildings).

Viewshed map generation Given a digital elevation model and the locations of one or more viewpoints, generate polygons enclosing the area visible from at least one viewpoint.

Virtual display A display management scheme that displays as much of an image as will fit on the monitor, while buffering the rest of the image in a temporary file from a RAM disk or the hard drive. A virtual display works like a window that can slide anywhere on a raster object.

Workstation A computer that consists of a graphic terminal, central processor, digitizer, graphics tablet (optional), and a mouse (optional). It may also be a stand-alone central processing unit (CPU) and its peripheral devices. It is often linked to other computers through a network.

Z-value Usually a reference to the value of surface height in a Digital Elevation/Terrain Model at a particular x,y location.

BIBLIOGRAPHY

Aangeenburg, R.T. *A Critique of GIS in Geographic Information Systems: Principles and Applications,* Maguire, D.J., Goodchild M.F., Rhind, D.W. (eds.). London: Longman Scientific and Technical. Vol. 1. 1991. 101-107.

Antenucci, J.C., Brown, K., Croswell, P.L., Kevany, M., Archer, H. *Geographic Information Systems: a guide to the technology*. Chapman and Hall. New York. 1991.

Asrar, G (ed.). *The theory and application of optical remote sensing*. New York, John Wiley & Sons Inc. 1989.

Baker, J. R., Mitchell P.L., Cordey, Groom, G. B., Settle, J. J., Stileman, M. R. Relationship between physical characteristics and polarmetric radar backscatter for Corsican pine stands in Thatford forest, UK. *International Journal of Remote sensing* 15. 1994. 2827-50.

Baker, W., Cai. Y., The r.le programs for multiscale analysis of landscape structure using the GRASS *geographical information system*. Landscape Ecology 7. 1992. 291-302.

Ballantyne, C.K., The Loch Lomond Readvance on the Isle of Skye, Scotland: glacier reconstruction and palaeoclimatic implications. *Journal of Quaterly Sci*ence, Vol.4, 2. 1989. 95-108.

Baltsavias, E.P. *Digital Ortho-Images - A Powerful Tool for the Extraction of Spatial and Geo-Information. PRS,* Vol. 51, (2). 1996. 63 – 77.

Barnsley, M.J. Environmental Monitoring using Multiple View Angle (MVA) Remotely-Sensed images in: *Environmental remote sensing from regional to global scales*. Foody, G., Curran, P.J. (eds.). London, Taylor and Francis. 1994. 181-201.

Barr, S. L., Barnsley, M.J., A region-based, graph-theoretic data model for the inference of second-order thematic information from remotely-sensed images. *International Journal of Geographical Information Science*, 11. 1997. 555-16.

Bernhardsen, T., Choosing a GIS in: Longley, P.A., Goodchild, M.F., Maguire, D.J., Rhind, D.W., (eds.). *Geographical Information Systems: Principles, Techniques, Management and Applications*. 2nd edn. Chichester: John Wiley. 1999. 589-600.

Bernhardsen, T. *Geographic Information Systems – An Introduction*. 2nd edn. 1999.

Berry, J. K. Beyond Mapping: Concepts, Algorithms, and Issues in GIS. Collins, F., *Colorado: GIS World*. 1993.

Berry, J.K. Fundamental operations in computer-assisted map analysis. *International Journal of GIS,* 1. 1987. 119-36.

Bezdek, J.C., Ehrlich, R., Full, W. FCM: the Fuzzy C-means clustering algorithm. *Computers and Geosciences* 10. 1984. 191-203.

Bill, Thoen., *History of Geographic Information Systems: Perspectives from the Pioneers*, Timothy, W. Foresman, (ed.), Prentice Hall PTR. 1998. 397.

Brusegard, D., Menger, G. Real Data and Real Problems: Dealing with large spatial databases in: Goodchild, M. and Gopal, S. (eds.) *Accuracy of Spatial Databases*. Taylor and Francis. 1989. 177-186.

Burrough, P.A., Mcdonnell, R.A. *Principles of Geographical Information Systems*. New York, Oxford University Press. 1998.

Burrough, P.A., Brown, L., Morris, E.C. Variations in vegetation and soil pattern across the Hawkesbury Sandstone plateau from Barren Grounds to Fitzroy Falls, New South Wales. *Australian Journal of Ecology*, 2. 1977. 137-59.

Burrough, P.A. Multiscale sources of spatial variation in soil (1). The application of fractal concepts to nested levels of soil narration. *Journal of Soil Science*, 34. 1983. 577-97.

Burrough, P.A., McDonnel R.A. *Principles of Geographical Information Systems for Land Resource Assessment*. Clarendon Press. Oxford. 1997.

Campbell, J. *Map Use and Analysis*. 2001. 37-47.

Carter, J.R. On defining the geographic information system in: W.J. Ripple (ed.), *Fundamentals of Geographical Information Systems: A Compendium*. ASPRS/ACSM, Falls Church. 1989. 3-7.

Carver, S.J., Cornelius, S.C., Heywood, D.I., Sear, D.A., Using Computers and Geographical Information Systems for Expedition Fieldwork. *The Geographical Journal*, Vol. 161, (2). 1995.167-176.

Chakraborty, Debashis., Dutta, D., Chandrasekharan, H. Land use indicators of a watershed in arid region, western Rajasthan using remote sensing and GIS. *Journal of Ind. Soc. Remote Sensing,* 29(3). 2001. 115-128.

Chakraborty, Debashis., Dutta, D., Chandrasekharan, H. Spatial modelling for hydrological response behaviour of an arid watershed, India – remote sensing and GIS approach. *Journal of Spatial Hydrology*, Spring, 5(1). 2004. 48-66.

Chakraborty, Debashis., Dutta, D., Chandrasekharan, H. Remote sensing application in spatial modelling of runoff in a watershed. *Indian Journal of Soil Cons.* 33(2). 2005. 110-113.

Chrisman, Nicholas. *Exploring Geographic Information Systems*. John Wiley and Sons. 1997.

Clarke, K.C. *Analytical and computer cartography*. Englewood Cliffs, NJ: Prentice-Hall. 1990.

Codd, E. F. A relational model of data for large shared data banks. *Communications of the ACM*, 13(6). 1970. 377–387.

Codd, E. F. Relational data base: A practical foundation for productivity. *Communications of the ACM,* 25(2). 1982. 109-117.

Collins, F. C., Bolstad, P. V. A Comparison of Spatial Interpolation Techniques in Temperature Estimation. *Proceedings CD rom.*Third, International Conference/Workshop on Integrating GIS and Environmental Modeling. Santa Fe, New Mexico, Jan 1996. 21-25.

Congalton, R., A review of assessment the accuracy of classifications of remotely sensed data. R*emote Sensing of Environment*, 37. 1991. 35-46.

Cooke, R.U., Doomkamp, J.C. *Geomorphology in Environmental Management- A New Introduction* 2nd edn., Clarendon Press, Oxford. 1990.

Craig, T. A. Treatise on Projections. Washington, DC: U.S. *Government Printing Office*. 1882.

Curan, P. *Principles of Remote Sensing*. Longman, London. 1989.

Curran, P.J. *Principles of Remote Sensing*. Longman, New York. 1985. 273

Dana, P. H. "Map Projections." http://www.colorado.edu/geography/gcraft /notes/mapproj/

Dangermond, J. A Classification of Software Components Used in Geographic Information Systems, Proc. US - Australia *Workshop on the Design and Implementation of Computer Based Geographic Information Systems*, Honolulu Hawaii. 1982. 70-91.

Davis, F. W., Simonett, D.S. GIS and remote sensing in: Maguire, D.J., Goodchild, M. F., Rhind, D. W. (eds.)., *Geographical information systems; principles and application*. Harlow, Longman/New York, John Wiley & Sons Inc. Vol. 1. 1991. 191-213.

Davis, B.E. Ch: Spatial Data, in: *GIS: A Visual Approach*, 1st edn, Santa Fe, NM, Onword Press. 1996.

Davis, J.C. Statistics and Data Analysis in: *Geology*, 2nd edn, Wiley, New York. 1986.

DeMers, Michael N. Classification and Purpose in Automated Vegetation Maps, *Journal of Geographical Review* 81(3). 1991. 267-280.

DeMers, Michael, N. *Fundamentals of Geographic Information Systems,* 2nd edn, John Wiley and Sons, New York. 2000. 498.

DeMers, Michael, N. GIS Modeling in Raster – I.S. (edn.), John Wiley and Sons, New York. 2001. 320.

Dent, B. *Cartography*: Thematic Map Design. 1999. 35-44.

Duerr, W.A., Teeguarden, D.E. *Forest resource management*: *Principles and Case Studies*. Philadelphia, WB, Sanders. 1979.

Ehlers, M., Edwards, G., Bedards, Y., Integration of remote sensing with geographical information systems: a necessary evolution. *Photogrammetric and Remote Sensing,* 55. 1989. 1619-27.

Ehlers, M., Multisensor image fusion techniques in remote sensing. ISPRS *Journal of Photogrammetry and Remote Sensing*, Vol. 46. 1991. 19-30.

Eitenbichler, S.B., (ed.), *GIS in Business* 93 *Conference Proceedings*. Boston: GIS World, Inc. 1993.

Environmental Systems Research Institute (ESRI). *Understanding GIS*: *The Arc/Info Method.* Redlands CA: ESRI. 1990.

Floyd, F., Sabins, *Journal of Remote Sensing*: *Principles and Interpretation*. Freeman, W.H. NY. 1997.

Foody, G.M. A Fuzzy sets approach to the representation of vegetation continua from remotely sensed data: an example from lowland heath. *Photogrammetric Engineering and Remote Sensing,* 58. 1992. 221-5.

Furuti, Carlos, A. Other Interesting Projections. http://www.progonos.com/ furuti /MapProj/ Normal/ ProjAz/ projAz.html. 1997.

Gahegan, Mark N., Stuart A. Roberts. An Intelligent, Object-Oriented Geographical Information System. *International Journal of Geographical Information Systems*. Vol. 2, (2) 1988. 101-110.

Goel. N.S. Inversion of Canopy reflectance models for estimation of biophysical parameters from reflectance data in: Asrar, G (ed.). *Theory and application of optical remote sensing*. New York, John Wiley and Sons Inc. 1989. 205-51.

Gong, P., Pu, R., Chen, J., Mapping ecological land systems and classification uncertainties from digital elevation and forest-cover data using neural networks. 1996.

Goodchild. Spatial Analysis and GIS. Pre-Conference Seminar, 2001 ESRI User Conference. *National Center for Geographic Information and Analysis*, University of California, Santa Barbara, CA93106.

Grandi, G., de., Lemoine, G.G., Groof. H., de., Lavalle, C., Siebert, A. J., Fully polarmetric classification of the Black Forest Maestro 1 Airsar data. *International Journal of Remote Sensing,* 15. 1994. 2755-76

Green, D. R. and King, S. D. (eds.)., *Coastal and Marine Geo-Information Systems: Applying the Technology to the Environment, Coastal Systems and Continental Margins*, Vol. 4. Kluwer, The Netherlands, 2003. 580.

Haykin, Simon. *Neural Networks, A Comprehensive Foundation.* New York: Macmillan College Publishing Company, 1994. 696.

Hepner, G.F., Logan, T., Ritter N., Bryant, N. Artificial Neural Network Classification Using a Minimal Training Set: Comparison to Conventional Supervised Classification, in: *Photogrammetric Engineering and Remote Sensing*, April. 1990. 469-473.

Hinton, J. C. GIS and remote sensing integration for environmental applications, *International Journal of Geographic Information System*, 10. 1996. 877-890.

Hinton, G. E., *Training Products of Experts by Maximizing Contrastive Likelihood.* Technical Report: GCNU TR1999001. Gatsby Computational Neuroscience Unit, University College London. 1999. (www.gatsby.ucl.ac.uk).

Hsu, K., Gupta, H., Sorooshian, S. 1995. link.aip.org/link/?JWRMD5/128/190/1.

Hsu, Sao-lin., Gupta H.V., Sorooshian, Soroosh. Artificil Neural Network modeling of the rainfall-runoff process. *Water Resources Research*, Vol. 31, (10). 1995. 2517-2530.

Hutchinson, C.F., Integration of remote sensing and farming systems research in West Africa, in: Bishay, A., Dregne, H.E. (eds.). *Advances in desert and arid land technology and development, Vol. 5,* (1). New York: Harwood Academic Publishers. 1991. 59-68.

Huxhold, W.E., *An Introduction of Urban Geographic Information Systems.* New York and Oxford: Oxford University Press. 1991.

Information technology Services, 2004. http://www.utexas.edu/its/windows/ database/datamodeling/rm/rm3.html

Richards, J.A., *Remote Sensing Digital Image Analysis*, Springer, Berlin. 1993.

Jacquemoud, S., Baret, F., Prospect: A Model of Leaf Optical Properties Spectra, *Remote Sensing of Environment*, 34. 1990. 251-256.

Janssen, L., Jaarsma, M., Londen, van der. Integrating topographic data with remote sensing for land cover classification, *Photogrammetric Engineering and Remote Sensing*, 5. 1990. 1503-1506.

Janssen, L.L.F., Middelkoop, H. Knowledge-based crop classification of a Landsat TM image. Int. *Journal of Remote Sensing*, 13. 1992. 2827-2837.

Jeffers, J.N.R. *An introduction to systems analysis with ecological application.* London, Edward Arnold. 1978.

Kandel, A. *Fuzzy mathematical techniques with applications.* Reading, MA: Addison-Wesley. xiv. 1986. 274.

Kauffmann, A. *Introduction to the Theory of Subsets*, Academic Press, New York. 1975.

Kemp, K., Vckovski, A. Towards an Ontology of Fields in *Third International Conference on GeoComputation*, Bristol, UK. 1998.

Klir, G.J., Folger, T.A. *Fuzzy sets, uncertainty, and information.* Englewood Cliffs, NJ: Prentice-Hall. xi. 1988. 355.

Krige, D.G., *Journal of Chem. Metall. Min. Soc. South Afr.* 52. 1951. 119-139.

Krige, D.G., *Journal of South Afr. Soc. Inst. Min. Metall.* 61. 1960. 231-233.

Lakhan, V.C., *Introductory Geographical Information Systems.* Summit Press, Toronto, 415.

Lam, N.S.M. Description and measurement of Landsat TM image using fractals. *Photogrammetric Engineering and Remote Sensing*, 56. 1990. 187-195.

Laurini, R., Thompson, D. *Fundamentals of spatial information systems*. Academic Press, London. 1992.

Lee, D., Sallee G. A method of measuring shape, *Geographical Review*, 60. 1970. 555-563.

LeGro. Assessing patch shape in landscape mosaics. *Photogrammetric Engineering and Remote Sensing*, 57. 1990. 285-293.

Li, X. Development of Neural Network Spatial Interpolator for Precipitation Estimation. *Proceedings of GIS/LIS'97 Annual Conference*, Cincinnati, Ohio. 1997. 653-667.

Goodchild, M.F. A spatial analytic perspective on geographical information systems. *International Journal of Geographical Information Systems*, 1. 1988. 327–334.

Maguire, D.J. An overview and definition of GIS, in: *Geographical Information Systems. Principles* (Maguire, D.J. et al.). Vol. 1: Longman, London, UK. 1991. 9-20

Maguire, David, Ch 21: The Functionality of GIS in: Maguire, D.J. Goodchild, M.F., Rhind, D.W. (eds.)., *Geographical Information Systems*: *Principles and Applications*. Longman, London. 1991.

Marble, D.F., Lauzon, J.P., McGranaghan, M. Development of a Conceptual Model of the Manual Digitizing Process, *Proceedings of the International Symposium on Spatial Data Handling*, Vol. 1, Aug 20-24, 1984, Zurich Switzerland, Symposium Secretariat, Department of Geography, University of Zurich-Irchel, 8057 Zurich, Switzerland. Conceptual discussion of the digitizing process. 1984.

Martin, D. *Geographical Information Systems and their socioeconomic applications*, 2nd edn. Routledge, London. 1996.

Mather, M.P. *Computer applications in geography*. Wiley, NY. 1991.

Matheron, G. *Principles of geostatistics*. Econ. Geol. 58. 1963. 1246-1266.

Matheron, G. *The theory of regionalized variables and their applications*. Ecole ds Mines, Fountainebleu, France. 1971.

McGarigal, K., Marks, B.J. Spatial pattern analysis program for quantifying landscape structure. *Dolores* (CO): P.O. Box 606. 1994. 67.

McHarg, I. L. *Design with Nature,* American Museum of Natural History, the Natural History Press, Garden City, N.Y. 1969.

Murai, Shunji. 1998. http://www.profc.udec.cl/~gabriel/tutoriales/giswb/vol1/author.htm

Muttiah, R. S., Allen, P. M. Srinivasan, R., Arnold, J. Baseflow mapping of the south eastern and south central United States using GIS in: *Application of GIS to the Modeling of Non-point Source Pollutants in the Vadose Zone*., Dennis L. Corwin., Keith Loague., (eds.)., Soil Science Society of America. SSSA Special Publication (48). 1996. 247-257.

Myeni, R.B., Williams, D.L. On the relationship between FPAR and NDVI. *Remote Sensing of Environment*, 49. 1994. 200-211.

Myneni, R. B., Williams, D.L. On the relationship between FAPAR and NDVI. *Remote Sensing of Environment*, 49. 1994. 200-211.

O'Sullivan, D. Too much of the wrong kind of data: implications for the practice of micro-scale spatial modelling in: Goodchild, M.F. and Janelle, D. (eds.)., *Spatially Integrated Social Science: Examples of Best Practice,* Oxford University Press: Oxford. 2004. 95–107.

Odum, E.P., Ecology, London, Holt Reinhart and Winston. 1975.

Pariente, D., Geographic Interpolation and Extrapolation by Means of Neural Networks. EGIS/MARI'94 *Proceedings*. Vol. 1. 1994. 684-693.

Pathan, S.K., Sastry, S.V.C., Dhinwa, P.S., Mand Majumdar, K.L.Urban growth trend analysis using GIS techniques – case study of the Bombay metropolitan region. *International Journal of Remote Sensing*, 14. 1993. 3169-3179.

Pavlidis, T. *Algorithms for Graphics and Image Processing*, Springer-Verlag, Berlin. 1982.

Peucker, T. K., Chrisman, N. Cartographic Data Structures, *American Cartographer*, Vol. 2, (1), 1975. 55-69.

Peuquet, D. J. A conceptual framework and comparison of spatial data models in: Peuquet D. J., Marble D.F. (eds.)., *Introductory Readings in Geographical Information Systems: Principles and Applications*. Taylor and Francis, London. 1990. 209-14.

Rees, W.G. *Physical Principles of Remote Sensing*. Cambridge (UK), Cambridge University Press. 1990.

Reitman, Valerie. Firms scrap plans to sell database with personal details on consumers. *The Philadelphia Inquirer* 1. Jan. 24, 1991.

Robinson, A.H., Morrison, J.L., Muehrcke, P.C., Kimerling, A.J., Guptill, S.C. *Elements of cartography*, 6th edn., John Wiley and Sons, NY. 1995.

Rosenfeld, A. *Quadtrees and pyramids in Proceedings of the 5th International Conference on Pattern Recognition.* Miami Beach, 1980. 802—811.

Sabins, Floyd, F. *Remote Sensing*: *Principles and Interpretation.* Freeman and Company, New York. 1999. 450.

Sahoo, R.N., Tomar, R.K., Abrol, I.P., Krishnamurty, V.V.R.S., Wadhwani, M.K., Tiwari, M.K., Drivers of land use dynamics in Gangetic plains: Application of Remote Sensing and GIS analysis. Proc. LUCC-ISPRS workshop "*LUCC Contribution to Asian Environmental Problems*, NRSA, Hyderabad, Dec. 2, 2002. 20-21.

Sahoo, R.N., Tomar, R.K., Sehegal, V.K., Rao, C.S., Nirupa Charchi., Abrol, I.P., Wadhwani, M.K., Gupta, R.K. Geo-informatics for enhancing crop productivity in eastern flood plains - A case study. Proc. Natl. Symp. on *Geo-informatics Applications for Sustainable Development* held at IARI, New Delhi, Feb. 17-19. 2004. 183-189.

Sárközy, F. The GIS concept and the three-dimensional modeling. *Computers, Environment and Urban Systems*. Vol. 18. (2), 1994. 111-121.

Smith T.R., Menon, S., Starr, J.L., Estes, J.E. Requirements and principles for the implementation and construction of large-scale geographic information systems. *International Journal of Geographical Information Systems*, 1. 1987. 13-31.

Snyder, John P. Map Projections—*A Working Manual*. U.S. Geological Survey Professional Paper:1395. 1987.

Star, J., Estes, J., *Geographic Information Systems*: *an Introduction*. Englewood Cliffs. Prentice Hall. 1990.

Star, J., Estes, J. *Geographic Information Systems-An Introduction*. Prentice Hall, New Jersey, USA. 1999.

Steinitz, C., Parker, P., Jordan, L. *Hand-drawn overlays: their history and prospective uses, Landscape Architecture*, Vol. 66, (5). 1976. 444-445.

Strahler, A.H., Woodcock, C.E., Smith, J.A., On the nature of models in remote sensing. *Remote sensing of environment*, 20. 1986. 121- 139.

Sullivan, John M. Stereographic Projection Demo. 2004. http://torus.math.uiuc.edu/jms/java/stereop/

Synder, John P. *Map projections: a working manual.* USGS Professional Paper: 1395. Washington, DC: United States Government Printing Office. 1987.

Thorp, P.W., A mountain icefield of Loch Lomond Stadial age, Western Grampians, Scotland. *Boreas*, Vol. 15. 1986. 83-97.

Tobler, W. R. Automation and Cartography. *Geographical Review*, Vol. 49. 1959. 526-534.

Tobler, W.R., Smooth pycnophylactic interpolation for geographical regions, *Journal of the American Statistical Association*. Sárközy F., and Gáspár P.1992. Modelling of scalar fields represented by scattered 3D points. *Periodica Polytechnica Civil Engineering*. Vol. 36, (2). 1992. 187-200.

Tomlin, Dana. *Geographic Information Systems and Cartographic Modeling*. Prentice Hall. 1990.

Torge, Wolfgang. *Geodesy*, 2nd edn., New York: de Gruyter. 1991

Townshend, J.R.G., Justice C.O. The spatial variation of vegetation changes at very coarse scales . *International Journal of Remote Sensing*, 11. 1990. 149-157.

Tsichritzis, D.C., Lochovsky, F.H. *Database Management Systems*, Academic Press, New York. 1977.

Tufte, E.R. *Envisioning Information.* Graphics Press, Cheshire, Conn. 1990.

Tufte, E.R., *The Visual Display of Quantitative Information*, Graphics Press, Cheshire, C.T. Contains numerous examples of graphical excellence (and its opposite) in map design. 1983.

Tyner, J. *Introduction to Thematic Cartography*, Ch 5: 75-109; Appendix C: Commonly Used Map Projections. 1992. 281-284.

Unesco. 1999. http://ioc.unesco.org/oceanteacher/resourcekit/Module2/GIS/Module/

van Roessel, J.W. Design of a spatial data structure using the relational normal forms, *International Journal of Geographical Information Systems* 1. 1987. 33-50.

Wahl, F. M. *Digital Image Signal Processing*. Boston and London: Artech. 1987.

Walker, J.D., Black, R.A., Linn, J.K., Thomas, A.J., Wiseman, R., D'Attilio, M.G., Development of Geographic Information Systems-Oriented Database for Integrated Geological and Geophysical Applications. *GSA Today: A Publication of the Geological Society of America* 6(3). 1996. 2-7.

Walsh, M.E. U.S. bioenergy crop economic analysis: status and needs. *Biomass and Bioenergy* 14(4). 1998. 341-350.

Weisstein, E. W. Books about Cartography. http://www.ericweisstein.com/ encyclopedias/books/ Cartography.html.

Weisstein, Eric W. Stereographic Projection. 2004. From *Math World*—A Wolfram Web Resource. http:/ /mathworld.wolfram.com/StereographicProjection.html.

Wilkinson, G. A review of Current Issues in the integration of GIS and Remote Sensing Data. *International Journal of Geographic Information Systems* 10. 1996. 85-101.

Woodsford, P. Spatial database update — a key to effective automation, *International Archives of Photogrammetry and Remote Sensing*, 31(4). 1996. 956–961.

Worboys, M.F., Hearnshaw, H. M., Maguire, D. J., Object-oriented data modelling for spatial databases. 1990. *International Journal of Geographical Information Systems*, 4(4). 1990. 369-383,

Worboys, M.F., Hearnshaw, H.M., Maguire, D.J. Object-oriented data and query modelling for geographical information systems, *Proceedings of the 4th International Symposium on Spatial Data Handling*, Zurich. 1990. 679-689.

Yang, W., Gold, Ch. Dynamic Spatial Object Condensation Based on the Voronoi Diagram. *Proceedings of the Fourth International Symposium* of LIESMARS titled: Toward three dimensional, temporal and dynamic spatial data modeling and analysis. Liesmars, Wtusm, P.R. China, Oct. 25-27. 1995. 134-145.

Yoshida, T., Omatu, S. Neural network approach to land cover mapping. *IEEE Transactions on Geoscience and Remote Sensing*, 32. 1994. 1103-1108.

Zadeh, L. A., Fuzzy sets, *Information and Control*, 8. 1965. 338-353.

Index